ANNALS *of* THE NEW YORK ACADEMY OF SCIENCES

T0179974

EDITOR-IN-CHIEF
Douglas Braaten

ASSOCIATE EDITOR
Rebecca E. Cooney

PROJECT MANAGER
Steven E. Bohall

EDITORIAL ADMINISTRATOR
Daniel J. Becker

Artwork and design by Ash Ayman Shairzay

The New York Academy of Sciences
7 World Trade Center
250 Greenwich Street, 40th Floor
New York, NY 10007-2157

annals@nyas.org
www.nyas.org/annals

**The New York
Academy of Sciences**

Published by Blackwell Publishing
On behalf of the New York Academy of Sciences

Boston, Massachusetts
2012

ANNALS *of* THE NEW YORK ACADEMY OF SCIENCES

VOLUME
1261

ISSUE

Neuroimmunomodulation in Health and Disease I

Basic Science

ISSUE EDITORS

Adriana del Rey,[a] C. Jane Welsh,[b] Markus J. Schwarz,[c] and Hugo O. Besedovsky[a]

[a]University of Marburg, [b]Texas A&M University, and [c]Ludwig-Maximilian University

TABLE OF CONTENTS

Academy Membership: Connecting you to the nexus of scientific innovation

Since 1817, the Academy has carried out its mission to bring together extraordinary people working at the frontiers of discovery. Members gain recognition by joining a thriving community of over 25,000 scientists. Academy members also access unique member benefits.

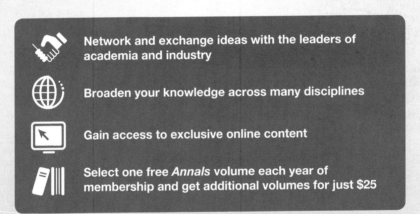

- **Network and exchange ideas with the leaders of academia and industry**
- **Broaden your knowledge across many disciplines**
- **Gain access to exclusive online content**
- **Select one free *Annals* volume each year of membership and get additional volumes for just $25**

Join or renew today at **www.nyas.org**.
Or by phone at **800.843.6927** (**212.298.8640** if outside the US).

Ann. N.Y. Acad. Sci. ISSN 0077-8923

ANNALS OF THE NEW YORK ACADEMY OF SCIENCES

Issue: *Neuroimmunomodulation in Health and Disease*

Foreword for *Neuroimmunomodulation in Health and Disease*

The immune system contributes to maintaining constancy of molecular and cellular components in an organism, and to facilitating adaptation not only to changeable environmental but also endogenous conditions. As in all physiological systems, the immune system is under the control of the neuroendocrine system and it interacts and works in concerted ways with other systems of the body. These interactions are based on the capacity of immune cells to produce chemical mediators—such as cytokines—capable of affecting the nervous and endocrine systems. Immune cells respond to neuroendocrine signals by way of specific receptors for hormones, neurotransmitters, and neuropeptides expressed on immune cells at different stages of their development and activation. This results in a neuroendocrine level of regulation that is interwoven with autoregulatory immune mechanisms. On the other hand, some immune products—for example, certain cytokines—are produced in the brain under basal conditions and, following increased neuronal and immune activity, contribute to modulate brain function. Furthermore, under particular conditions, immune, neural, and endocrine cells produce the same mediators and share the use of second messengers, protein kinases, transcription factors, and other intracellular mediators, and regulate posttranscriptional events. Since the immune, endocrine, and nervous systems are always active and their signals are produced at multiple cellular and organ levels, terms such as *bidirectional communication, feed-back,* and *reflexes* have become too restricted; we now realize that neuro–endocrine–immune interactions can only be considered as dynamic networks that tend to reach an adaptive equilibrium during health and disease. The growing evidence that stimuli that originally were thought only to elicit an immune response can also trigger neural and endocrine responses, and that those that were thought only to influence the brain and associated mechanisms can also exert immunoregulatory effects, indicates the expanded operation of this network. This network of interactions also mediates metabolic adjustments that can influence both immune and brain functions. As with all physiological mechanisms, disruptions, distortions, or imbalances in the network may lead to disease.

This and an accompanying volume of *Annals of the New York Academy of Sciences* present contributions on the main topics discussed at the 8th Congress of the International Society for Neuroimmunomodulation (ISNIM), held on October 20–22, 2011, and organized with the German Endocrine-Brain-Immune Network in Dresden, Germany. The oral presentations at the congress were selected to provide a wide spectrum of research devoted to interactions among the immune, endocrine, and nervous systems. As indicated by the name chosen for this congress, Neuroimmunomodulation in Health and Disease: An Integrative Biomedical Approach, speakers discussed not only basic research regarding neuro–endocrine–immune interactions but also the potential relevance of research for clinical practice. Consistent with this aim, the contributions

doi: 10.1111/j.1749-6632.2012.06677.x

presented in both *Annals* volumes have been divided into the basic and translational aspects of neuroimmunomodulation.

The first volume, *Neuroimmunomodulation in Health and Disease I: Basic Science*, covers selected molecular mechanisms underlying neuro–endocrine–immune interactions, the relevance of these interactions for the development of the immune system and thymic function, and the establishment of self-tolerance. Examples of papers in this section include those describing how immune cells can influence the innervation of lymphoid organs and how neurotransmitters and hormones affect immune and inflammatory processes. The possible implications of complex neuroendocrine responses for immunoregulation during stress, sleep, and aging, and also during the course of the body's circadian rhythms, are also included in this section.

The second volume, *Neuroimmunomodulation in Health and Disease II: Translational Science*, includes studies dealing with infectious diseases, such as tuberculosis, HIV, and parasitic diseases. These studies are followed by reports related to neuropsychiatric disorders such as schizophrenia, depression, and Alzheimer's disease. This volume also presents examples of neuro–immune–endocrine interactions during pain, and in autoimmune, inflammatory, and neoplasic diseases.

We thank the authors for their valuable contributions and the reviewers for their helpful comments. And while we regret that not all of the excellent oral and poster presentations at the congress could be included in these volumes, we trust that the selected studies are representative examples of the intense efforts currently being made to provide further relevance for integrative views of biology, physiology, and medicine.

ADRIANA DEL REY
University of Marburg

C. JANE WELSH
Texas A&M University

MARKUS J. SCHWARZ
Ludwig-Maximilian University

HUGO O. BESEDOVSKY
University of Marburg

Ann. N.Y. Acad. Sci. ISSN 0077-8923

ANNALS OF THE NEW YORK ACADEMY OF SCIENCES

Issue: *Neuroimmunomodulation in Health and Disease*

Expression and functions of μ-opioid receptors and cannabinoid receptors type 1 in T lymphocytes

Jürgen Kraus

Department of Pharmacology and Toxicology, University of Magdeburg, Magdeburg, Germany

Address for correspondence: Jürgen Kraus, Department of Pharmacology and Toxicology, University of Magdeburg, 44 Leipzigerstrasse, 39120 Magdeburg, Germany. juergen.kraus@med.ovgu.de

Opioids and cannabinoids modulate T lymphocyte functions. Many effects of the drugs are mediated by μ-opioid receptor and cannabinoid receptor type 1 (CB1), respectively. These two receptors are strikingly similar with respect to their expression in T cells and the mechanisms by which they mediate modulation of T cell activity. Thus, μ-opioid receptors and CB1 not expressed in resting primary human and Jurkat T cells. However, in response to the cytokine IL-4, the epigenetic modifiers 5-aza-2′-deoxycytidine and trichostatin A, and activation of T cells, functional μ-opioid receptors and CB1 are induced. The induced receptors mediate inhibition of T cell signaling and, thereby, IL-2 production, a hallmark of activated T cells. Although coupled to inhibitory G proteins, μ-opioid receptors and CB1 produce a remarkable increase in cAMP levels in T cells stimulated with opioids and cannabinoids, which is a key mechanism for the inhibition of T cell signaling.

Keywords: μ-opioid receptors; CB1; opioid; cannabinoid; T lymphocytes; immunomodulation

Introduction

Opioids and cannabinoids are best known for affecting neuronal function. The effects of most of the commonly used opioids like morphine are mediated by μ-opioid receptors.[1] Among these effects, analgesia is clinically of utmost importance. The cannabinoid receptor type 1 (CB1) is one of the most abundant receptors in the brain and is involved in, for example, emotion and reward.[2,3] In addition to neuronal effects, opioids and cannabinoids mediate a large number of similar, immunomodulatory effects, in particular, effects on T lymphocytes.[4–6] To mention only two examples, both groups of drugs regulate the T helper (Th) cell balance and induce a shift toward Th2 cells.[7–9] In addition, opioids and cannabinoids inhibit the production of IL-2 of activated T cells, which is a hallmark of activated T cells.[10–15]

The expression of specific receptors is a prerequisite for effects of certain ligands on certain cells. Although there has been plenty of indirect evidence for the expression of μ-opioid receptors on T cells derived from such *in vivo* data as mentioned previously, and although transgenic mice lacking the μ-opioid receptor gene have obvious defects in proper T cell functions,[16,17] it has been a matter of debate for a long time, whether μ-opioid receptors, identical to those found in neuronal cells, were expressed in T cells at all. One of the first key steps toward the understanding of this issue was the discovery in my laboratory in 2001 that the expression of μ-opioid receptors in primary human T cells, and also in all other immune effector cells that were tested, is induced *de novo* in response to the cytokine IL-4.[18] With respect to cannabinoid receptors, the existence of a central/neuronal receptor (CB1) and a second peripheral/non-neuronal receptor (CB2) was an accepted paradigm for many years.[19] Today this view has changed, because inducible expression,[20,21] as well as specific effects[15] of CB1 in T cells, have been demonstrated by several investigators.

In my laboratory, we are focusing on the expression and functions of μ-opioid receptors and CB1 in human T lymphocytes. Over the last years, we have observed a striking similarity between the two receptors in this respect, which will be summarized

doi: 10.1111/j.1749-6632.2012.06524.x

in this report. In particular, both receptors are induced in T cells by IL-4, by the epigenetic modifiers 5-aza-2′-deoxycytidine (5-Aza-dC), which is an inhibitor of DNA methyltransferases, and by trichostatin A (TSA), which is a histone deacetylase inhibitor, and in response to activation of T cells. With respect to similar effects in T cells, we have shown that µ-opioid receptors and CB1 mediate the same key mechanisms, which are the basis for the inhibition of the production of IL-2 in activated T cells by opioids and cannabinoids.

It should be mentioned that T cells are functionally divided into different subsets, for example, Th1, Th2, Th17, Th23, or regulatory T cells, and that effects of opioids and cannabinoids may be different in the different subsets. However, at the moment there is too little detailed information on the exact phenotype of the T cells investigated. Therefore, the general term *T cells* is used throughout this paper.

Induction of µ-opioid receptors and CB1 in response to IL-4

In cultured primary human T cells and in cells of the human T cell line Jurkat, the expression of µ-opioid receptors and CB1 is repressed. However, the expression of both receptors is markedly induced in response to IL-4. The genes are transactivated by the transcription factor STAT6, which typically mediates effects of IL-4. It has been demonstrated that STAT6 binds to nucleotide −997 of the µ-opioid receptor promoter[18,22–25] and nucleotide −2769 of the CB1 promoter.[21,26]

Interestingly, µ-opioid receptors and CB1 not only are induced by IL-4, but also promote the induction of IL-4 in T lymphocytes. Thus, it has been reported that activation of µ-opioid receptors, for example, by morphine, results in an induction of IL-4 in T cells.[25,27] Likewise, treatment of T cells with cannabinoids results in induction of IL-4.[24,28] In this way, a positive feedback loop is generated that is likely to be a basic mechanism for the modulation of the Th cell balance by opioids and cannabinoids. In addition, one should expect cross-induction of both µ-opioid receptors and CB1 by both opioids and cannabinoids via the common mediator IL-4. Such a scenario might explain the multiple interactions between opioids and cannabinoids, which are observed at several levels, for example, in behavioral studies. However, until now, such cross-induction on a molecular level has not been reported.

The positive feedback mechanism involving µ-opioid receptors, CB1, and IL-4 may also explain, in part, anti-inflammatory effects of opioids and cannabinoids. Although anti-inflammatory effects of IL-4—the prototypical anti-inflammatory cytokine[29]—are well recognized, there is increasing evidence for such effects of opioids[30,31] and cannabinoids.[32,33] Anti-inflammatory effects of these drugs also include the inhibition of proinflammatory mediators like the cytokine TNF-α and its primary mediator, the transcription factor NF-κB. Such effects have been observed for opioids (Börner *et al.*) as well as cannabinoids.[34,35] In conclusion, both drugs should be beneficial in anti-inflammatory therapies in humans. In fact, a combination of cannabidiol and tetrahydrocannabinol (termed nabiximols) is already used as an additional drug for the treatment of multiple sclerosis.

Induction of µ-opioid receptors and CB1 in response to epigenetic factors

Correct temporal and spatial gene expression requires regulatory mechanisms at several levels. Epigenetic gene regulation contributes to such mechanisms. It has been shown that agents that modify the epigenetic architecture of a cell may induce the expression of certain genes. In particular, hypomethylation of a given gene, especially of its promoter (e.g., with inhibitors of DNA methyltransferases like 5-Aza-dC), often results in the transcriptional activation of the gene. Similar effects are observed after hyperacetylation of residues of the histones in which a gene is packed, for example, with histone deacetylase inhibitors like TSA. Often, a combination of such drugs has synergistic effects on the induction of genes.[36–38] We recently reported that treatment of Jurkat cells with 5-Aza-dC and TSA, either alone or in combination, results in the induction of µ-opioid receptor– and CB1-specific mRNA and functional receptor protein (Börner *et al.*). Of note, treatment of SH SY5Y neuroblastoma cells, which constitutively express µ-opioid receptor and CB1 genes, with 5-Aza-dC and TSA did not result in altered expression. This indicates that 5-Aza-dC and TSA modulate the expression of genes that are epigenetically repressed, rather than the expression of genes that are not repressed. Interestingly, similar but opposite observations were made for CB2. Although expressed constitutively in Jurkat T cells, this receptor is repressed in SH SY5Y cells. In response to 5-Aza-dC and TSA, CB2 expression is induced in SH SY5Y cells but not in Jurkat cells.

There is an increasing number of pharmacologically used drugs that act by altering the epigenetic code. For example, in the treatment of certain tumors it has been demonstrated that reactivation of the expression of tumor suppressor genes by certain drugs has beneficial effects.[39,40] Generally, elevated expression of a certain receptor may result in increased effects of cognate ligands/drugs. To this end, increased opioid and cannabinoid receptor expression in T lymphocytes could be useful for increasing anti-inflammatory effects of certain drugs, and thus increase their therapeutical potentials.

It should be mentioned that epigenetic mechanisms are also involved in the induction of μ-opioid receptors by IL-4.[41] Our observations have indicated that several modifications in histones and the μ-opioid receptor promoter DNA occur within the first hours after stimulation of T cells with IL-4. After these modifications, binding of activated STAT6 to the cis-active −997 element could be detected.

Induction of μ-opioid receptors and CB1 in response to activation of T lymphocytes

In resting T cells, signaling of the T cell receptor complex is tonically inhibited by phosphorylation of the inhibitory regulatory site Tyr505 of the kinase Lck. Activation of the cells by ligation of the T cell receptor (e.g., by antigen-presenting cells) results in a loss of the inhibitory effects of Lck, which itself then phosphorylates and activates the kinase Zap70. This induces a signaling cascade that involves several kinases and calcium and culminates in the activation of the three transcription factors AP-1, NF-κB, and NFAT.[42–44] One of the prominent effects of the activation of this cascade is the induction of IL-2, which induces well-defined subsequent immune responses.[29] In addition to IL-2, expression of μ-opioid receptors[45] and CB1[20] are induced.

Thus, it has been shown that activation of primary human T cells, as well as Jurkat cells, results in induction of functional μ-opioid receptors.[46] Using decoy oligonucleotides directed against AP-1, NF-κB, and NFAT, it has been demonstrated that all of the three factors are involved in this effect. In addition, functional cis-active binding sites for AP-1 (−2388 and −1434 (Ref. 46)), NF-κB (−2174, −557, and −207 (Ref. 47)), and NFAT (−1064, −785,

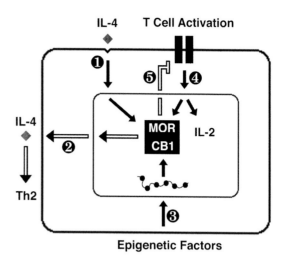

Figure 1. Expression and functions of μ-opioid receptors and CB1 in T lymphocytes. The expression of of μ-opioid receptors (MOR) and CB1 in T cells is induced by the cytokine IL-4 (1). Mediated by these receptors, opioids and cannabinoids induce the expression of IL-4, which promotes differentiation into Th2 cells (2). The expression of μ-opioid receptors and CB1 is also induced by the epigenetic modifyers 5-Aza-dC and TSA (3). Typically, T cell activation results in the induction of IL-2 (4). In addition, μ-opioid receptors and CB1 are induced (D). Via these receptors, opioids and cannabinoids inhibit T cell signaling (5).

and −486 (Ref. 45)) have been characterized and localized on the human μ-opioid receptor promoter. Similarly, CB1 is induced in response to activation of primary human T cells and Jurkat cells. Experiments with decoy oligonucleotides have also demonstrated involvement of AP-1, NF-κB, and NFAT.[20] However, experimental evidence has yet to localize the binding sites for these TFFs on the human CB1 promoter.

Inhibition of the production of IL-2 by opioids and cannabinoids in activated T cells

It has been known for some decades that opioids[10,11] and cannabinoids[13,14] inhibit the production of IL-2 in activated T cells. This effect may be beneficial in diseases like multiple sclerosis, where "overshooting" activities of T cells are involved. On the other hand, this effect may also explain immunosuppressive properties of opioids and cannabinoids. Recently, the molecular mechanisms underlying the inhibition of IL-2 by opioids[12] and cannabinoids[15] have been elucidated. Strikingly, the key event in the mechanistic chain resulting in the inhibition of T cell signaling—namely, a remarkable increase of intracellular cAMP levels—is identical for both drugs. The elevated cAMP levels then inhibit the signaling cascade of activated T cells in a PKA-dependent mechanism by which the inhibitory effect of the kinase Lck, which exerts a tonic inhibition on this cascade,[42,44] is further enhanced. In this way, initiation of T cell signaling, culminating in the production of IL-2, is inhibited. This is a striking finding because μ-opioid receptors and CB1, in addition to CB2, that mediate this effect (as demonstrated by specific agonists and antagonists) are coupled to inhibitory G_i proteins. Such receptors are classically associated with decreased cAMP levels. In fact, incubation of T cells with opioids or cannabinoids results in decreased cAMP levels, though only within a short, initial period lasting only for several minutes. This is followed by a drastic and long-lasting increase in cAMP production. Thus, opioids produce cAMP levels that exceed control levels by approximately sixfold and last for several hours. At the 24-h time point, opioid-induced cAMP levels are again similar to control levels because of the induction of phosphodiesterases.[12] For cannabinoids, this effect is even stronger; without inducing phosphodiesterases, they produce cAMP levels that exceed control levels up to 10-fold and persist over a period of several days.[15] The reason for this switch, from inhibition to facilitation of cAMP production, is largely unknown. It may be speculated that other (or additional) adenylyl cyclases are recruited or that the activity of these enzymes is altered upon prolonged ligand binding and receptor activation.

Conclusions and perspectives

G protein–coupled receptors are found in most higher organisms and are evolutionarily very old. However, the lack of conserved sequences makes it unclear whether the members of this superfamily have derived from a single, common ancestral gene.[48] At the moment, it is also not clear whether the similarities between μ-opioid receptors and CB1, with respect to their expression and functions in T cells, reflect evolutionary mechanisms (Fig. 1 summarizes features about the expression and function of μ-opioid receptors and CB1 in T lymphocytes). There has likely been a coevolution of parts of the genes—for example, in promoter regions—that determines their expression profiles. With respect to the expression of μ-opioid receptors and CB1 in T cells, it should be noted that repression of the genes

is characteristic for resting primary T cells and T cells in culture under *in vitro* conditions. *In vivo*, however, there is reason to believe that μ-opioid receptors and CB1 are expressed at high levels in T cells because of the multiple mechanisms that induce their expression. In fact, the addition of human serum of different donors to resting primary human T cells in culture can induce μ-opioid receptors and CB1 (Kraus, unpublished observation). This hypothesis is further strengthened by the multiple reports demonstrating various effects of opioids and cannabinoids on T-cell function *in vivo*. It will be interesting to identify additional similarities between μ-opioid receptors and CB1. Preliminary observations from my laboratory indicate that the CB1 gene in T cells is also induced by the cytokine TNF-α via the transcription factor NF-κB, which, again, would be similar to μ-opioid receptors, for which such regulation has already been reported.[47,49]

Acknowledgments

The author wishes to thank Christine Börner for helpful discussion and critical review of this manuscript.

Conflicts of interest

The author declares no conflicts of interest.

References

1. Kieffer, B.L. & C.J. Evans. 2009. Opioid receptors: from binding sites to visible molecules *in vivo. Neuropharmacology* **56** (Suppl 1): 205–212.
2. Di Marzo, V. 2008. CB(1) receptor antagonism: biological basis for metabolic effects. *Drug Discov. Today* **13**: 1026–1041.
3. Lutz, B. 2009. Endocannabinoid signals in the control of emotion. *Curr. Opin. Pharmacol.* **9**: 46–52.
4. Sacerdote, P. 2008. Opioid-induced immunosuppression. *Curr. Opin. Support. Palliat. Care* **2**: 14–18.
5. Roy, S. *et al.* 2006. Modulation of immune function by morphine: implications for susceptibility to infection. *J. Neuroimmune Pharmacol.* **1**: 77–89.
6. Bisogno, T. & V. Di Marzo. 2010. Cannabinoid receptors and endocannabinoids: role in neuroinflammatory and neurodegenerative disorders. *CNS Neurol. Disord. Drug Targets* **5**: 564–573
7. Roy, S. *et al.* 2004. Chronic morphine treatment differentiates T helper cells to Th2 effector cells by modulating transcription factors GATA 3 and T-bet. *J. Neuroimmunol.* **147**: 78–81.
8. Sacerdote, P. *et al.* 1998. Endogenous opioids modulate allograft rejection time in mice: possible relation with Th1/Th2 cytokines. *Clin. Exp. Immunol.* **113**: 465–469.

9. Klein, T.W. *et al.* 2004. Cannabinoid receptors and T helper cells. *J. Neuroimmunol.* **147:** 91–94.

10. Roy, S. *et al.* 1997. Morphine inhibits transcriptional activation of IL-2 in mouse thymocytes. *Cell. Immunol.* **179:** 1–9.

11. Thomas, P.T., R.V. House & H.N. Bhargava. 1995. Direct cellular immunomodulation produced by diacetylmorphine (heroin) or methadone. *Gen. Pharmacol.* **26:** 123–130.

12. Börner, C. *et al.* 2009. Mechanisms of opioid-mediated inhibition of human T cell receptor signaling. *J. Immunol.* **183:** 882–889.

13. Kaplan, B.L. *et al.* 2005. Inhibition of leukocyte function and interleukin-2 gene expression by 2-methylarachidonyl-(2′-fluoroethyl)amide, a stable congener of the endogenous cannabinoid receptor ligand anandamide. *Toxicol. Appl. Pharmacol.* **205:** 107–115.

14. Klein, T.W. *et al.* 1995. Delta 9-tetrahydrocannabinol, cytokines, and immunity to legionella pneumophila. *Proc. Soc. Exp. Biol. Med.* **209:** 205–212.

15. Börner, C. *et al.* 2009. Cannabinoid receptor type 1- and 2-mediated increase in cyclic AMP inhibits T cell receptor-triggered signaling. *J. Biol. Chem.* **284:** 35450–35460.

16. Gaveriaux-Ruff, C. *et al.* 1998. Abolition of morphine-immunosuppression in mice lacking the mu-opioid receptor gene. *Proc. Natl. Acad. Sci. USA* **95:** 6326–6330.

17. Roy, S., R.A. Barke & H.H. Loh. 1998. MU-opioid receptor-knockout mice: role of mu-opioid receptor in morphine mediated immune functions. *Brain Res. Mol. Brain Res.* **61:** 190–194.

18. Kraus, J. *et al.* 2001. Regulation of mu-opioid receptor gene transcription by interleukin-4 and influence of an allelic variation within a STAT6 transcription factor binding site. *J. Biol. Chem.* **276:** 43901–43908.

19. Howlett, A.C. *et al.* 2002. International Union of Pharmacology. XXVII. Classification of cannabinoid receptors. *Pharmacol. Rev.* **54:** 161–202.

20. Börner, C., V. Hollt & J. Kraus. 2007. Activation of human T cells induces upregulation of cannabinoid receptor type 1 transcription. *Neuroimmunomodulation* **14:** 281–286.

21. Börner, C. *et al.* 2008. Analysis of promoter regions regulating basal and interleukin-4-inducible expression of the human CB1 receptor gene in T lymphocytes. *Mol. Pharmacol.* **73:** 1013–1019.

22. Kraus, J., C. Borner & V. Hollt. 2003. Distinct palindromic extensions of the 5′-TTC...GAA-3′ motif allow STAT6 binding *in vivo*. *FASEB J.* **17:** 304–306.

23. Börner, C. *et al.* 2004. STAT6 transcription factor binding sites with mismatches within the canonical 5′-TTC...GAA-3′ motif involved in regulation of delta- and mu-opioid receptors. *J. Neurochem.* **91:** 1493–1500.

24. Börner, C., V. Hollt & J. Kraus. 2006. Cannabinoid receptor type 2 agonists induce transcription of the μ-opioid receptor gene in jurkat T cells. *Mol. Pharmacol.* **69:** 1486–1491.

25. Börner, C. *et al.* 2007. Comparative analysis of mu-opioid receptor expression in immune and neuronal cells. *J. Neuroimmunol.* **188:** 56–63.

26. Börner, C. *et al.* 2007. Transcriptional regulation of the cannabinoid receptor type 1 gene in T cells by cannabinoids. *J. Leukoc. Biol.* **81:** 336–343.

27. Roy, S. *et al.* 2005. Morphine induces CD4+ T cell IL-4 expression through an adenylyl cyclase mechanism independent of the protein kinase A pathway. *J. Immunol.* **175:** 6361–6367.

28. Yuan, M. *et al.* 2002. Delta 9-Tetrahydrocannabinol regulates Th1/Th2 cytokine balance in activated human T cells. *J. Neuroimmunol.* **133:** 124–131.

29. Curfs, J.H., J.F. Meis & J.A. Hoogkamp-Korstanje. 1997. A primer on cytokines: sources, receptors, effects, and inducers. *Clin. Microbiol. Rev.* **10:** 742–780.

30. Takeba, Y. *et al.* 2001. Endorphin and enkephalin ameliorate excessive synovial cell functions in patients with rheumatoid arthritis. *J. Rheumatol.* **28:** 2176–2183.

31. Philippe, D. *et al.* 2003. Anti-inflammatory properties of the mu opioid receptor support its use in the treatment of colon inflammation. *J. Clin. Invest.* **111:** 1329–1338.

32. Molina-Holgado, F. *et al.* 2003. Endogenous interleukin-1 receptor antagonist mediates anti-inflammatory and neuroprotective actions of cannabinoids in neurons and glia. *J. Neurosci.* **23:** 6470–6474.

33. Nakajima, Y. *et al.* 2006. Endocannabinoid, anandamide in gingival tissue regulates the periodontal inflammation through NF-κB pathway inhibition. *FEBS Lett.* **580:** 613–619.

34. Panikashvili, D. *et al.* 2005. CB1 cannabinoid receptors are involved in neuroprotection via NF-κ B inhibition. *J. Cereb. Blood Flow Metab.* **25:** 477–484.

35. Mormina, M.E. *et al.* 2006. Cannabinoid signalling in TNF-alpha induced IL-8 release. *Eur. J. Pharmacol.* **540:** 183–190.

36. Berger, S.L. 2007. The complex language of chromatin regulation during transcription. *Nature* **447:** 407–412.

37. Jenuwein, T. & C.D. Allis. 2001. Translating the histone code. *Science* **293:** 1074–1080.

38. Wolffe, A.P. & M.A. Matzke. 1999. Epigenetics: regulation through repression. *Science* **286:** 481–486.

39. Karpf, A.R. & D.A. Jones. 2002. Reactivating the expression of methylation silenced genes in human cancer. *Oncogene* **21:** 5496–5503.

40. Jones, P.A. 1986. DNA methylation and cancer. *Cancer Res.* **46:** 461–466.

41. Kraus, J. *et al.* 2010. Epigenetic mechanisms involved in the induction of the mu opioid receptor gene in Jurkat T cells in response to interleukin-4. *Mol. Immunol.* **48:** 257–263.

42. Koretzky, G.A. & P.S. Myung. 2001. Positive and negative regulation of T-cell activation by adaptor proteins. *Nat. Rev. Immunol.* **1:** 95–107.

43. Horejsi, V., W. Zhang & B. Schraven. 2004. Transmembrane adaptor proteins: organizers of immunoreceptor signalling. *Nat. Rev. Immunol.* **4:** 603–616.

44. Torgersen, K.M. *et al.* 2001. Release from tonic inhibition of T cell activation through transient displacement of C-terminal Src kinase (Csk) from lipid rafts. *J. Biol. Chem.* **276:** 29313–29318.

45. Börner, C. *et al.* 2008. T cell receptor/CD28-mediated activation of human T lymphocytes induces expression of functional {micro}-opioid receptors. *Mol. Pharmacol.* **74:** 496–504.

46. Börner, C., V. Hollt & J. Kraus. 2002. Involvement of activator protein-1 in transcriptional regulation of the

human mu-opioid receptor gene. *Mol. Pharmacol.* **61:** 800–805.

47. Kraus, J. *et al.* 2003. The Role of nuclear factor {κ}B in tumor necrosis factor-regulated transcription of the human {micro}-opioid receptor gene. *Mol. Pharmacol.* **64:** 876–884.

48. Graul, R.C. & W. Sadee. 2001. Evolutionary relationships among G protein-coupled receptors using a clustered database approach. *AAPS PharmSci.* **3:** E12.

49. Beltran, J.A., A. Pallur & S.L. Chang. 2006. HIV-1 gp120 up-regulation of the mu opioid receptor in TPA-differentiated HL-60 cells. *Int. Immunopharmacol.* **6:** 1459–1467.

Ann. N.Y. Acad. Sci. ISSN 0077-8923

ANNALS OF THE NEW YORK ACADEMY OF SCIENCES
Issue: *Neuroimmunomodulation in Health and Disease*

Reconciling neuronally and nonneuronally derived acetylcholine in the regulation of immune function

Koichiro Kawashima,[1,2] Takeshi Fujii,[3] Yasuhiro Moriwaki,[4] Hidemi Misawa,[4] and Kazuhide Horiguchi[5]

[1]Department of Molecular Pharmacology, Kitasato University School of Pharmacy, Tokyo, Japan. [2]Research Institute of Pharmaceutical Sciences, Musashino University, Tokyo, Japan. [3]Department of Pharmacology, Doshisha Women's College of Liberal Arts, Kyotanabe, Kyoto, Japan. [4]Department of Pharmacology, Keio University, Tokyo, Japan. [5]Department of Anatomy, University of Fukui, Matsuoka, Fukui, Japan

Address for correspondence: Koichiro Kawashima, Ph.D., Department of Molecular Pharmacology, Kitasato University School of Pharmacy Tokyo 108-8641, Japan. koichiro-jk@piano.ocn.ne.jp; kkawashimak@pharm.kitasato-u.ac.jp

Immune cells, including lymphocytes, express muscarinic and nicotinic acetylcholine (ACh) receptors (mAChRs and nAChRs, respectively), and agonist stimulation of these AChRs causes functional and biochemical changes in the cells. The origin of the ACh that acts on immune cell AChRs has remained unclear until recently, however. In 1995, we identified choline acetyltransferase mRNA and protein in human T cells, and found that immunological T cell activation potentiated lymphocytic cholinergic transmission by increasing ACh synthesis and AChR expression. We also found that M_1/M_5 mAChR signaling upregulates IgG_1 and proinflammatory cytokine production, whereas $\alpha7$ nAChR signaling has the opposite effect. These findings suggest that ACh synthesized by T cells acts as an autocrine and/or paracrine factor via AChRs on immune cells to modulate immune function. In addition, a recently discovered endogenous allosteric $\alpha7$ nAChR ligand, SLURP-1, also appears to be involved in modulating normal T cell function.

Keywords: T cell; choline acetyltransferase; cytokine; mAChR; nAChR; SLURP-1

Introduction

In 1914, acetylcholine (ACh) was isolated from the ergot produced by the fungus *Claviceps purpurea* as a compound that exerted inhibitory effects on the heart and stimulatory effects on intestinal smooth muscle.[1] Later, in 1929, Dale and Dudley[2] accidentally discovered the presence of ACh in the spleens of the ox and horse. This was the first occasion on which ACh was found to occur naturally in an animal, prompting Dale to recognize ACh as the transmitter for the vagus nerve,[3] which complimented Loewi's discoveries relating to chemical transmission of nerve impulses in the isolated frog heart.[4] Since then, ACh has been generally known as an important neurotransmitter in the nervous systems of vertebrates and insects; however, both ACh and ACh-synthesizing activity have also been detected in most life forms without nervous systems, including plants, fungi, and even bacteria and archea.[5,6] What is more, ACh has been detected in a number of non-neuronal cells from mammalian species, including lymphocytes and vascular endothelial cells (reviewed in Refs. 7–13). Collectively, these findings suggest that ACh has been present from the very beginnings of life, and that, in addition to its role as a neurotransmitter, it serves as a local mediator regulating a variety of other physiological functions (reviewed in Refs. 5, 6, 8–10, 12, 13).

Within the cholinergic neuron, ACh is synthesized by choline acetyltransferase (ChAT) from choline and acetyl coenzyme A.[14] Once released from nerve terminals, ACh is hydrolyzed within a few milliseconds into choline and acetate by acetylcholinesterase (AChE) and butyrylcholinesterase (BuChE), which are found within the

doi: 10.1111/j.1749-6632.2012.06516.x
Ann. N.Y. Acad. Sci. 1261 (2012) 7–17 © 2012 New York Academy of Sciences.

synaptic cleft and in circulation. Immune cells, such as lymphocytes, dendritic cells, and macrophages, express both muscarinic and nicotinic ACh receptors (mAChRs and nAChRs, respectively; reviewed in Refs. 7–11, 15), stimulation of which by their respective agonists causes functional and biochemical changes that include increased cell proliferation and increases in the intracellular free calcium ion concentration ($[Ca^{2+}]_i$). Given, however, the high catalytic activities of AChE and BuChE and the lack of evidence for direct contact between cholinergic nerve endings and target immune cells, the origin of the ACh that would act via AChRs on immune cells had long been an enigma. However, using a sensitive and specific radioimmunoassay, we were able to detect a substantial amount of ACh in the blood of various animals, and subsequently detected ChAT-catalyzed ACh synthesis in CD4[+] T cells (reviewed in Refs. 7–9, 15). Since then, we have focused on characterizing the contribution made by this apparent lymphocytic cholinergic system to the regulation of immune function (reviewed in Refs. 7–11).

On the basis of the observation that vagus nerve stimulation protected mice from lethal endotoxic shock induced by lipoplysaccharide (LPS),[16] Pavlov *et al.*[17] proposed the presence of a cholinergic anti-inflammatory pathway in which parasympathetic cholinergic neurons provide hard-wired innervation to macrophages; ACh released from cholinergic nerve endings through vagus stimulation would act via α7 nAChRs on the macrophages, inhibiting the synthesis and release of tumor necrosis factor (TNF-α), thereby preventing lethal shock. But although the spleen is essential for the cholinergic antiinflammatory pathway,[18] the vagal cholinergic nerve terminates in the celiac superior mesenteric plexus, and no cholinergic neurons reach the spleen.[19,20] Recently, Rosas-Ballina *et al.*[21] found that ACh synthesized by a subset of CD4[+] T cells that express ChAT plays a key role in inhibiting TNF-α production by splenic macrophages during vagus nerve stimulation. These findings confirm our initial observation that among the T cell subsets, CD4[+] T cells exhibit ChAT-catalyzed ACh synthesis,[22,23] and they support the notion that ACh released from T cells triggers nuclear signaling via mAChRs and nAChRs expressed on their own plasma membrane or on other cells located in the vicinity; that is, they

act as autocrine and/or paracrine factors (reviewed in Refs. 7–9, 11, 15).

This review focuses primarily on the presence of ACh in the the blood, its synthesis by ChAT in T cells, and the regulatory mechanisms governing its synthesis and release in immune cells, expression of AChRs in immune cells; roles of AChRs in the regulation of immune function; and the prospective roles of the recently identified endogenous allosteric α7 nAChR ligand, secreted lymphocyte antigen-6/urokinase-type plasminogen activator receptor-related peptide-1 (SLURP-1), in the regulation of immune function.

Presence of ACh in the blood, ACh synthesis by ChAT in T cells, and regulation of the synthesis and release of ACh in immune cells

ACh in the blood

In 1930, Kapfhammer and Bischoff[24] reported for the first time the presence of ACh in the blood of oxen. Thereafter, there were a number of attempts to detect ACh in the blood of a variety of animal species, but because of the unavailability of sensitive and reliable methods for assaying ACh, as well as the enzymatic and physicochemical instability of ACh, for a long period there were no consistent data available to confirm the presence of ACh in blood. In 1989, however, Kawashima *et al.* began using a sensitive and specific radioimmunoassay to convincingly demonstrate the presence of ACh in the blood of a number of mammalian species (Table 1).[25–29] Furthermore, after fractionation of human blood into plasma, mononuclear leukocytes (MNLs), polymorphonuclear leukocytes (PNLs), and red blood cells, they found that about 60% of the total blood ACh content (1,263.5 ± 149.0, mean ± SEM, $n = 30$) was present in the MNL fraction (731.0 ± 85.8 pg/ml), which was composed mainly of lymphocytes and a small number of monocytes.[26] No ACh was detected in the PNL fraction, composed mainly of neutrophils, or in the red cell fraction, suggesting blood ACh originates mainly from lymphocytes. ChAT activity was also detected in the MNL fraction of rabbits and humans.[30,31]

ACh synthesis by ChAT in T cells

In 1995, Fujii *et al.*[22] provided the first definitive evidence that ACh synthesis in T cells is catalyzed by

Table 1. ACh concentrations in the blood of various mammalian species

Species	Blood ACh (pg/mL)
Bovine	
Holstein (7)[a]	$52,636 \pm 8,718$
Holstein (5)[b]	$62,212 \pm 2,683$
Japanese black (8)	$51,733 \pm 2,649$
Chimpanzee (6)	$3,143 \pm 380$
Dog, beagle (7)	451 ± 65
Goat (5)	592 ± 136
Horse (5)	$13,685 \pm 2,384$
Human (30)	$1,264 \pm 149$
Porcine (5)	$1,701 \pm 230$
Rabbit, Japanese white (7)	$3,722 \pm 897$
Rat (10)	209 ± 29
Sheep (5)	298 ± 29

NOTE: Values are means $\pm$ SEM. Numbers of samples included are shown in parentheses. Statistics a arranged from data presented in Refs. 25–29.

[a]Samples obtained from a ranch affiliated with the Animal Husbandry Experimental Station, Department of Agriculture, University of Tokyo.

[b]Samples obtained from a ranch at Nisshin Flour Milling Co., Ltd.

ChAT. They demonstrated that both ChAT mRNA and protein were expressed in MOLT-3 cells, a human leukemic T cell line (Fig. 1A). Later, ChAT mRNA was also detected in peripheral blood MNLs from humans and animals,[32,33] confirming that lymphocytes have the capacity for ACh synthesis catalyzed by ChAT. CD4[+] T cells account for the majority of ACh synthesis by ChAT (see Ref. 23; Fig. 1B), though expression of ChAT activity, mRNA, and protein has also been detected in CD8[+] T cells, dendritic cells, granulocytes, macrophages, and mast cells.[32–34]

Experiments carried out using cultured bovine arterial endothelial cells (ECs) and porcine brain microvessel ECs showed that in addition to immune cells, vascular ECs are also able to synthesize and release ACh.[35,36] This suggests the possibility that ACh released from ECs may interact with mAChRs on the same ECs to stimulate nitric oxide synthesis, and/or it may interact with AChRs on lymphocytes or macrophages during cell-to-cell interactions via intercellular adhesion molecules (ICAMs; reviewed in Refs. 8 and 9). Later, expression of immunoreactive ChAT protein was detected in human umbilical vein ECs, and ChAT mRNA expression was detected in pulmonary arterial ECs, confirming the synthesis of ACh by ChAT in vascular ECs.[37,38]

Regulation of the synthesis and release of ACh in immune cells

The regulation of ChAT expression has mainly been studied in T cells.[39] In 1996, Fujii *et al.*[40] showed that immunological activation of MOLT-3 and HSB-2 cells, two human leukemic T cell lines, by phytohemagglutinin (PHA) via T cell receptor (TCR)/CD3-mediated pathways increases both intracellular ACh content and its release. This suggests that PHA upregulates ACh synthesis in these cells. Later, it was confirmed that PHA and concanavalin A each upregulate ChAT mRNA expression in T cell lines and in MNLs isolated from human and animal blood.[15,23,41,42] It therefore seems likely that the interaction of T cells with dendritic cells during antigen presentation stimulates ACh synthesis in the T cells. Among the T cell subsets, CD4[+] T cells have a greater capacity to synthesize ACh than do CD8[+] T cells (Fig. 1B).[23,32]

Lymphocyte function–associated antigen-1 (LFA-1; CD11a/CD18) expressed on T cells plays a key role in leukocyte migration and T cell activation. Stimulation of CD11a (the LFA-1 α chain) in MOLT-3 cells using a CD11a monoclonal antibody (mAb) increases ChAT activity, the synthesis and release of ACh, and the expression of ChAT and M_5 mAChR mRNA.[44] This suggests that cell-to-cell adhesion of T cells to B cells, to antigen-presenting cells, or to ECs activates intracellular signaling pathways that lead to upregulation of lymphocytic cholinergic activity. Notably, simvastatin, a cholesterol-lowering drug, inhibits LFA-1 signaling by binding to an allosteric site on CD11a, suggesting that simvastatin exerts its immunosuppressive effects in part by modifying lymphocytic cholinergic activity.[44] In addition, calcium ionophores, the nonspecific protein kinase C (PKC) activator PMA, and the protein kinase A (PKA) activator dibutyryl cAMP all upregulate ChAT expression and ACh release from T cells (reviewed in Refs. 23, 39, and 45). Thus, the non-neuronal cholinergic system in immune cells also appears to be regulated by PKC, PKA, and changes in $[Ca^{2+}]_i$.

A ChAT mRNA and protein expression in MOLT-3

1. ChAT mRNA

2. ChAT protein

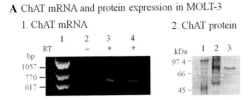

B ChAT mRNA expression in human T cells

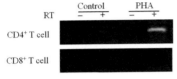

Figure 1. Expression of choline acetyltransferase (ChAT) mRNA and protein in human immune cells. (A1) Expression of ChAT mRNA in the MOLT-3 human leukemic T cell line. RT, reverse transcriptase. Lane 1, DNA size markers (bp); lane 2, MOLT-3 mRNA without RT (negative control); lane 3, MOLT-3 mRNA; lane 4, human brain mRNA (positive control). Arranged from data presented in Ref. 22. (A2) Western blot analysis of ChAT protein expression in MOLT-3 cells. Lane 1, molecular size markers (kDa); lane 2, human placental ChAT (positive control); lane 3, MOLT-3. (B) Expression of ChAT mRNA in human CD4[+] T cells and its potentiation by immunological activation with phytohemagglutinin (PHA). Note that CD8[+] T cells do not express ChAT mRNA, even after immunological activation. Arranged from data presented in Ref. 23.

Mechanism of ACh release

The mechanisms underlying ACh storage and release in the lymphocytic cholinergic system remain uncertain. In the nervous system, ACh is stored in synaptic vesicles through the action of vesicular acetylcholine transporter (VAChT) and released through exocytosis. Whether or not lymphocytic ACh is stored in vesicles has not yet been determined. Fujii *et al.*[41] could not confirm the presence of VAChT expression in human MNLs, which suggests that the mechanisms of ACh storage and release in the lymphocytic cholinergic system differ from those in the nervous system, though ACh storage for later release in as yet unidentified vesicles has not been completely ruled out (reviewed in Refs. 7–9). One possible alternative is that lymphocytic ACh is synthesized when necessary and then directly released without storage. In addition, Fujii *et al.*[46] recently detected expression of mediatophore mRNA and protein in the CCRF-CEM and MOLT-3 human leukemic T cell lines. Mediatophore is a homo-oligomer of a 16-kDa subunit homologous to the proteolipid subunit c of vacuolar H[+]-ATPase (V-ATPase), which is involved

in ACh release at the Torpedo nerve–electroplaque junction.[47] Notably, PHA-induced immunological activation of T cells upregulated mediatophore gene expression, and transfection of the cells with small interference RNA targeting mediatophore downregulated the protein and significantly reduced ACh release. These results suggest that ACh release from T cells is mediated, at least in part, via mediatophore.

AChRs expressed in immune cells

mAChRs

Five mAChR subtypes (M_1–M_5) that signal via two distinctly different second messenger systems have been identified.[48,49] The M_1, M_3, and M_5 mAChR subtypes are coupled to $G_{q/11}$, which, upon stimulation, mediates activation of phospholipase C, leading to increases in $[Ca^{2+}]_i$. The M_2 and M_4 mAChR subtypes are coupled to $G_{i/o}$, which, upon stimulation, mediates inhibition of adenylyl cyclase, leading to decreases in cAMP synthesis.[50] Most immune cells, including MNLs, dendritic cells, and macrophages, express all five mAChR subtypes (reviewed in Refs. 7–9, 15, and 51; Table 2); however, the expression level of each subtype appears to vary among animal species and with immune status.[53]

Stimulation of T cell mAChRs using ACh or an agonist such as bethanechol, carbachol, or oxotremorine-M (Oxo-M, an M_1/M_3 mAChR agonist) in the presence of extracellular Ca^{2+} causes a rapid increase in $[Ca^{2+}]_i$ followed by oscillations in $[Ca^{2+}]_i$ that persist for at least 10 minutes.[54] These effects are blocked by atropine, a nonspecific mAChR antagonist. In addition, Oxo-M upregulates expression of c-fos, an intranuclear transcriptional regulator,[55] and augments PHA-induced interleukin (IL)-2 production by upregulating IL-2 transcription in human lymphocytes.[56] These findings suggest that ACh released from T cells acts as an autocrine factor triggering nuclear signaling via mAChRs expressed on the T cell plasma membrane.

nAChRs

nAChRs exist as pentamers comprised of one to four distinct subunits (α, β, γ, δ, and ϵ) forming ligand-gated ion channels. Most nAChRs in nonneuronal cells are heteromeric pentamers comprised of $\alpha 2$–$\alpha 6$, $\beta 2$–$\beta 4$, and $\alpha 9/\alpha 10$ subunits, or they may be homomeric pentamers of the $\alpha 7$ subunit.[7–9] Agonist stimulation of most nAChRs causes membrane

Table 2. Expression of mRNAs encoding mAChRs and nAChRs in human immune cells

A. Expression of mRNAs encoding mAChR subtypes

Sample	Cell type	M_1	M_2	M_3	M_4	M_5
1 (F)	MNLs	+	+	+	+	+
2 (F)	MNLs	−	+	−	+	+
3 (F)	MNLs	+	+	+	+	+
4 (F)	MNLs	+	−	+	+	+
5 (M)	MNLs	+	+	−	+	+
6 (M)	MNLs	+	−	+	+	+
7 (M)	MNLs	−	+	+	+	+

B. Expression of mRNAs encoding nAChR subtypes

Sample	Cell type	α3	α5	α7	α9	α10
1 (F)	T	+	+	+	+	+
2 (F)	T	+	+	−	+	−
3 (F)	T	+	+	−	−	−
4 (F)	T	+	+	+	−	−
5 (F)	T	+	+	+	+	−
6 (F)	T	−	+	−	+	−
7 (F)	T	+	+	+	+	+
8 (F)	T	−	+	+	+	+
1 (F)	B	+	+	+	−	+
2 (F)	B	+	+	+	+	+
3 (F)	B	+	+	+	−	−
4 (F)	B	+	+	+	+	+
5 (F)	B	+	−	−	+	−
6 (F)	B	+	+	+	+	+
7 (F)	B	−	+	+	+	+
8 (F)	B	+	+	+	+	+

Arranged from the data presented in Ref. 52.
+, positive expression; −, negative expression; F, female; M, male; MNLs, mononuclear leukocytes.

depolarization and excitation due to a rapid increase in the membrane permeability to Na^+, K^+, and Ca^{2+}, though stimulation of nAChRs comprised of α9/α10 subunits leads to desensitization.[57]

Human T and B cells, leukemic cell lines, and thymocytes, as well as animal immune cells, all express various neuronal-type nAChR subunits (reviewed in Refs 7–9, 15, 51, and 52; Table 2). For example, Qian *et al.*[53] recently reported induction of α4 and α7 expression in murine $CD4^+$ T cells after activation of TCR/CD3-mediated pathways, which suggests that the pattern of nAChR subunit expression may reflect immune status. However, because of the abundance of nAChR subtypes, further study

will be required to determine whether the pattern of nAChR subunit expression varies within an individual depending on their immune status and/or genetic predisposition. Furthermore, the pattern of nAChR subunit expression in immune cells may vary among mammalian species.

Stimulation of the CCRF-CEM human leukemic T cell line using nicotine (0.01–10 μM) downregulates transcription of the α3, α5, α6, α7, and β4 nAChR subunits in a concentration-dependent manner.[58] In line with that observation, we found that levels of α5 and α7 subunit mRNA were significantly diminished in peripheral MNLs from smokers, compared to nonsmokers.[11] In the presence of extracellular Ca^{2+}, nicotine evoked rapid, transient increases in $[Ca^{2+}]_i$ in CCRF-CEM cells; this effect was inhibited by α-bungarotoxin or methyllycaconitine, two $\alpha7$ nAChR subunit antagonists, indicating that the $\alpha7$ nAChR subunit mediates Ca^{2+} signaling in T cells.[58] This suggests that nicotine, and therefore smoking, likely affects immune function by suppressing expression of various nAChR subunits involved in Ca^{2+} signaling in lymphocytes.

Role of AChRs in the modulation of immune function

mAChRs

When lymphocytes from M_1 mAChR knockout (KO) mice were stimulated *in vitro*, they exhibited a defect in their ability to differentiate into cytolytic $CD8^+$ T cells, though M_1 mAChRs do not appear to be required for early activation.[59] However, no defect was observed when expansion of $CD8^+$ T cells in M_1 or M_5 mAChR-KO mice was stimulated by infection with lymphocytic choriomeningitis virus or vesicular stomatitis virus, which suggests that M_1 and M_5 mAChRs are not involved in antiviral immunity.[60] On the other hand, keyhole limpet hemocyanin, a protein antigen that induces both B cell– and T cell–mediated responses, upregulates the activities of both ChAT and AChE in mouse lacrimal gland,[61] suggesting that protein antigens do stimulate lymphocytic cholinergic activity. Furthermore, T cell activation induced by PHA, PMA (which activates PKC- and mitogen-associated protein kinase pathways), or anti-CD11a mAb upregulated both M_5 mAChR and ChAT gene expression (reviewed in Refs. 23, 45, and 62). Stimulation of B cells in the MNL fraction using *Staphylococcus aureus* Cowan I also upregulated M_5 mAChR gene expression.[45] These findings suggest that immunological activation of lymphocytes facilitates cholinergic transmission by increasing M_5 mAChR expression and ACh synthesis.

The roles played by M_1 and M_5 mAChRs in the regulation of immune function were further investigated using M_1 and M_5 mAChR double KO (M_1/M_5 KO) mice immunized with ovalbumin (OVA).[63] Serum anti-OVA–specific IgG_1 concentrations determined one week after immunization were significantly lower in M_1/M_5 KO mice than in wild-type (WT) mice (Fig. 2A, 1a), but there was no difference in the serum anti-OVA–specific IgM concentration

between the two groups (Fig. 2A, 1b). When splenic MNLs consisting of 90% lymphocytes (30% $CD4^+$ and 15% $CD8^+$ T cells; 45% $CD45R^+$ B cells) and 10% monocytes ($CD14^+$ cells) prepared at the time of serum sampling were incubated with OVA, secretion of the proinflammatory cytokines TNF-α (Fig. 2A, 2c), interferon (INF)-γ (Fig. 2A, 2d), and IL-6 (Fig. 2A, 2e) from the activated MNLs was significantly diminished in M_1/M_5 KO mice, compared with WT mice. In addition, levels of AChE mRNA, an indicator of T cell activation, were lower in M_1/M_5 KO mice than WT mice.[63] These results support the notion that immune function is regulated, at least in part, via lymphocytic mAChR signaling, and that M_1 and/or M_5 mAChRs are involved in regulating proinflammatory cytokine production, leading to modulation of antibody production. Finally, Vira *et al.*[64] recently reported that M_3 mAChR KO mice infected with the gastrointestinal nematode parasite *Nippostrongylus brasiliensis* show a profound defect in the development of $CD4^+$ T cells, which correlates with protective immunity to the parasite. This suggests M_3 mAChRs play a key role in regulating $CD4^+$ T cell–driven Th2 immunity.

nAChRs

Stimulation of $\alpha7$ nAChRs on macrophages negatively regulates the synthesis and release of TNF-α, thereby protecting mice from lethal endotoxic shock induced by LPS.[65] The roles played by $\alpha7$ nAChRs in the regulation of immune function were investigated in nAChR $\alpha7$ subunit KO ($\alpha7$ KO) mice immunized with OVA.[66] Serum concentrations of anti-OVA–specific IgG_1 determined two weeks after immunization were significantly elevated in $\alpha7$ KO mice (Fig. 2B, 1a), whereas serum anti-OVA–specific IgM concentrations tended to be higher than in WT mice, but the difference was not significant (Fig. 2B, 1b). In addition, OVA-activated splenic MNLs from $\alpha7$ KO mice produced more TNF-α (Fig. 2B, 2c), IFN-γ (Fig. 2B, 2d), and IL-6 (Fig. 2B, 2e) than those from WT mice. These results suggest that $\alpha7$ nAChRs are involved in regulating production of proinflammatory cytokines, through which they modulate antibody production. Consistent with this idea, Skok *et al.*[67] reported the involvement of $\alpha7$ nAChRs, as well as other nAChR subunits, in the regulation of B cell development and activation.

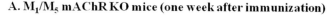

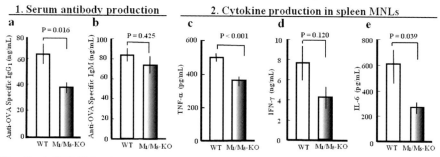

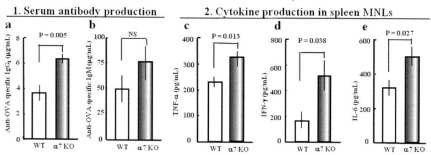

Figure 2. Contribution made by acetylcholine receptors to the regulation of immune function. (A1) Serum concentrations of anti-OVA–specific IgG_1 (1a) and IgM (1b) in M_1/M_5 mAChR-KO ($n = 9$) and WT ($n = 10$) C57BL/6-JJcl mice one week after immunization with OVA. In all panels in the figure, bars represent means ± SEM. (A2) TNF-α (2c), INF-γ (2d), and IL-6 (2e) production in splenic MNLs isolated from M_1/M_5-KO and WT mice immunized with OVA. Arranged from the data presented in Ref. 63. (B) Serum concentration of anti-OVA–specific IgG_1 (1a) and IgM (1b) in $\alpha7$ nAChR knockout ($\alpha7$ KO) ($n = 7$) and WT ($n = 7$) C57BL/6-J mice two weeks after immunization with OVA. (B2) TNF-α (2c), INF-γ (2d), and IL-6 (2e) production in splenic MNLs isolated from $\alpha7$ KO and WT mice immunized with OVA. Arranged from the data presented in Ref. 66.

Prospective roles of SLURP-1 in the regulation of immune function

SLURP-1 was first purified from human blood by Adermann *et al.*,[68] who then identified the protein using a urine peptide library. SLURP-1 acts as a positive allosteric ligand that potentiates the action of ACh at $\alpha7$ nAChRs[69,70] and stimulating proapoptotic activity in human keratinocytes.[70] Mutations in the gene encoding SLURP-1 have been detected in patients with Mal de Meleda (MDM), a rare autosomal recessive skin disorder characterized by transgressive palmoplantar keratoderma.[71] In addition to keratinocytes, SLURP-1 mRNA is expressed in most murine organs, including the thymus and spleen,[72] as well as various immune cells, including MNLs, dendritic cells, and macrophages.[15,72] Immunoreactive SLURP-1 has also been detected in human skin, keratinocytes, vaginal mucosal cells, and gingival cells, and in human colon cancer epithelial cells.[70,73–75] Interestingly, when Tjiu *et al.*[76] used

MNLs isolated from MDM patients to investigate the contribution made by SLURP-1 to the regulation of T cell function, they found a defect in the proliferative response to activation in T cells from MDM patients. Moreover, addition of WT recombinant SLURP-1 to cultures of T cells from MDM patients restored the normal T cell activation response. These findings show that by enhancing the actions of ACh mediated via $\alpha7$ nAChR signaling in T cells, SLURP-1 may play a key role during normal activation of T cells elicited by immunological stimulation.

More than 80% of vagus nerve fibers are afferent sensory neurons[20] that express peptide transmitters including substance P (SP)[77] and calcitonin gene-related peptide (CGRP).[78] Moriwaki *et al.*[79] found that immunoreactive SLURP-1 colocalizes with subsets of sensory neurons containing SP and CGRP. Furthermore, in the white pulp of the spleen, SP-positive nerve fibers have been identified in the marginal zone and in the outer regions

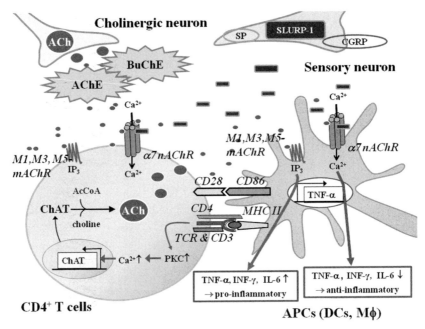

Figure 3. Schematic diagram illustrating our current understanding of the lymphocytic cholinergic system. $CD4^+$ T cells have a capacity for ChAT-catalyzed ACh synthesis that is activated during interaction with antigen-presenting cells (e.g., dendritic cells) and macrophages, and through cell-to-cell adhesion. ACh released from $CD4^+$ cells into the microenvironment during the interaction acts back on mAChRs and nAChRs expressed on the releasing cells' plasma membrane and on adjacent cells before it is hydrolyzed by AChE and BuChE. Stimulation of M_1/M_5 mAChRs upregulates proinflammatory cytokine production, leading to elevations in serum IgG_1. By contrast, $\alpha7$ nAChR signaling downregulates proinflammatory cytokine production, leading to a reduction of serum IgG_1. A newly identified endogenous $\alpha7$ nAChR allosteric ligand, SLURP-1, colocalizes with subsets of SP and CGRP neurons and potentiates the action of ACh at $\alpha7$ nAChRs. ACh, acetylcholine; AcCoA, acetyl coenzyme A; AChE, acetylcholinesterase; APC, antigen-presenting cell; BuChE, butyrylcholinesterase; ChAT, choline acetyltransferase; CGRP, calcitonin gene-related peptide; DCs, dendritic cells; IP_3, inositol-1,4,5-trisphosphate; mAChR, muscarinic ACh receptor; MHC, major histocompatibility complex; Mφ, macrophages; nAChR, nicotinic ACh receptor; PKC, protein kinase C; SP, substance P; TCR, T cell receptor.

of the periarteriolar lymphoid sheaths, which are filled with T cells. Recently, Horiguchi *et al.* identified immunoreactive SLURP-1 in large cytoplasm-rich splenic cells and neuron-like structures in the marginal zone, where they are surrounded by F4/80-positive macrophages and T cells (Horiguchi *et al.* in preparation). SLURP-1 has also been detected in human plasma, urine, sweat, saliva, and tears, suggesting that the molecule is fairly stable.[80] All of these findings are consistent with the idea that SLURP-1 is liberated antidromically from SP- and CGRP-containing neurons during vagus nerve stimulation, and from large cytoplasm-rich splenic cells by immune stimulation. Taken together, these findings suggest SLURP-1 participates in the modulation of immune function by acting as an allosteric ligand that potentiates the action of ACh on $\alpha7$ nAChRs in both T cells and macrophages (Fig. 3).

Conclusion

Immune cells, including lymphocytes, dendritic cells, monocytes, and macrophages, express all five mAChR (M_1–M_5) subtypes and various nAChR subunits ($\alpha2$–$\alpha7$, $\alpha9$–$\alpha10$, $\beta2$, and $\beta4$). Stimulation of these mAChRs and nAChRs by their respective agonists causes functional and biochemical changes in the immune cells. The possibility that ACh derived from parasympathetic cholinergic nerve terminals acts on immune cell AChRs would seem unlikely, as no evidence for direct cholinergic innervation of immune cells has yet been found. Among immune cells, it is mainly $CD4^+$ T cells that exhibit ChAT-catalyzed ACh synthesis. Moreover, immunological activation of T cells via TCR/CD3- or CD11a-mediated pathways upregulates ChAT and M_5 mAChR mRNA expression, suggesting regulation of lymphocytic cholinergic system by immunological stimulation. Studies with mAChR and nAChR KO mice revealed that M_1/M_5 mAChRs upregulate the production of TNF-α, IFN-γ, and IL-6 in immune cells, thereby elevating serum IgG$_1$, and that $\alpha7$ nAChRs downregulate the production of these same proinflammatory cytokine productions, thereby reducing serum IgG$_1$. These findings also demonstrate that, acting via AChRs, ACh released from T cells is involved in regulating cytokine production in immune cells, thereby modulating antibody production. Finally, a recently discovered endogenous $\alpha7$ nAChR allosteric ligand, SLURP-1, appears to be involved in modulating immune function by potentiating lymphocytic cholinergic activity.

Acknowledgments

Supported in part by funding from SSR Foundation and a generous donation from Dr. Eun Bang Lee, emeritus professor, College of Pharmacy, Seoul National University, Korea.

Conflicts of interest

The authors declare no conflicts of interest.

References

1. Ewins, A.J. 1914. Acetylcholine, a new active principle of ergot. *Biochem. J.* **8:** 44–49.
2. Dale, H.H & H.W. Dudley. 1929. The presence of histamine and acetylcholine in the spleen of the ox and the horse. *J. Physiol. (London)* **68:** 97–123.
3. Dale, H.H. 1935. Walter Ernest Dixon Memorial Lecture. Pharmacology and nerve-endings. *Proc. R. Soc. Med.* **28:** 15–28.
4. Loewi, O. 1921. Über humorale Übertrgbarkeit der Herzn-wirkung. *Pflügers Arch. Ges. Physiol.* **189:** 239–242.
5. Horiuchi, Y., R. Kimura, N. Kato, *et al.* 2003. Evolutional study on acetylcholine expression. *Life Sci.* **72:** 1745–1756.
6. Yamada, T., T. Fujii, T. Kanai, *et al.* 2005. Expression of acetylcholine (ACh) and ACh-synthesizing activity in archaea. *Life Sci.* **77:** 1935–1944.
7. Kawashima, K. & T. Fujii. 2000. Extraneuronal cholinergic system in lymphocytes. *Pharmacol. Ther.* **86:** 29–48.
8. Kawashima, K. & T. Fujii. 2003. Minireview: the lymphocytic cholinergic system and its contribution to the regulation of immune activity. *Life Sci.* **74:** 675–696.
9. Kawashima, K. & T. Fujii. 2004. Expression of non-neuronal acetylcholine in lymphocytes and its contribution to the regulation of immune function. *Front. Biosci.* **9:** 2063–2085.
10. Kawashima, K. & T. Fujii. 2008. Basic and clinical aspects of non-neuronal acetylcholine: overview of non-neuronal cholinergic systems and their biological significance. *J. Pharmacol. Sci.* **106:** 167–173.
11. Fujii, T., Y. Takada-Takatori & K. Kawashima. 2008. Basic and clinical aspects of non-neuronal acetylcholine: expression of an independent, non-neuronal cholinergic system in lymphocytes and its clinical significance in immunotherapy. *J. Pharmacol. Sci.* **106:** 186–192.
12. Grando, S.A., K. Kawashima, C.J. Kirkpatrick & I. Wessler. 2007. Introduction: recent progress in understanding the non-neuronal cholinergic system in humans. *Life Sci.* **80:** 2181–2185.
13. Wessler, I., C.J. Kirkpatrick & K. aRacké. 1998. Non-neuronal acetylcholine, a locally acting molecule widely distributed in biological system: expression and function in humans. *Pharmacol. Ther.* **77:** 59–79.
14. Tucek, S. 1982. The synthesis of acetylcholine in skeletal muscles of the rat. *J. Physiol. (London)* **322:** 53–69.
15. Kawashima, K., K. Yoshikawa, Y.X. Fujii, *et al.* 2007. Expression and function of genes encoding cholinergic components in murine immune cells. *Life Sci.* **80:** 2314–2319.
16. Borovikova, L.V., S. Ivanova, M. Zhang, *et al.* 2000. Vagus nerve stimulation attenuates the systemic inflammatory response to endotoxin. *Nature* **405:** 458–462.
17. Pavlov, V.A., H. Wang, C.J. Czura, *et al.* 2003. The cholinergic anti-inflammatory pathway: a missing link in neuroimmunomodulation. *Mol. Med.* **9:** 125–134.
18. Huston, J.M., M. Ochani, M. Rosas-Ballina, *et al.* 2006. Splenectomy inactivates the cholinergic antiinflammatory pathway during lethal endotoxemia and polymicrobial sepsis. *J. Exp. Med.* **203:** 1623–1628.
19. Bellinger, D.L., D. Lorton, R.W. Hamill, *et al.* 1993. Acetylcholinesterase staining and choline acetyltransferase activity in the young adult rat spleen: lack of evidence for cholinergic innervation. *Brain Behav. Immun.* **17:** 191–204.
20. Berthoud, H.R. & T.L. Powley. 1996. Interaction between parasympathetic and sympathetic nerves in prevertebral ganglia: morphological evidence for vagal efferent innervation of ganglion cells in the rat. *Microsc. Res. Tech.* **35:** 80–86.
21. Rosas-Ballina, M., P.S. Olofsson, M. Ochani, *et al.* 2011. Acetylcholine-synthesizing T cells relay neural signals in a vagus nerve circuit. *Science* **334:** 98–101.
22. Fujii, T., S. Yamada, H. Misawa, *et al.* 1995. Expression of choline acetyltransferase mRNA and protein in T-lymphocytes. *Proc. Japan Acad.* **71B:** 231–235.
23. Fujii, T., Y. Watanabe, K. Fujimoto & K. Kawashima. 2003. Expression of acetylcholine in lymphocytes and modulation of an independent lymphocytic cholinergic activity by immunological stimulation. *Biog. Amine.* **17:** 373–386.
24. Kapfhammer, J. & C. Bischoff. 1930. Acetylcholin und Cholin aus tierischen Organen. *Z. Physiol. Chem.* **191:** 179–182.
25. Kawashima, K., H. Oohata, T. Suzuki & K. Fujimoto. 1989. Extraneuronal localization of acetylcholine and its release upon nicotine stimulation. *Neurosci. Lett.* **104:** 336–339.
26. Kawashima, K., K. Kajiyama, K. Fujimoto, *et al.* 1993. Presence of acetylcholine in human blood and its localization in circulating mononuclear leukocytes. *Biog. Amine.* **9:** 251–258.
27. Fujii, T., S. Yamada, N. Yamaguchi, *et al.* 1995. Species differences in the acetylcholine content in blood and plasma. *Neurosci. Lett.* **201:** 207–210.

28. Yamada, S., T. Fujii, K. Fujimoto, *et al.* 1997. Oral administration of KW-5092, a novel gastroprokinetic agent with acetylcholinesterase inhibitory and acetylcholine release enhancing activities, causes a dose-dependent increase in the blood acetylcholine content of beagle dogs. *Neurosci. Lett.* **225:** 25–28.

29. Fujii, T., Y. Mori, T. Tominaga, *et al.* 1997. Maintenance of constant blood acetylcholine content before and after feeding in young chimpanzees. *Neurosci. Lett.* **227:** 21–24.

30. Kajiyama, K., T. Suzuki, K. Fujimoto & K. Kawashima. 1991. Determination of acetylcholine content and choline acetyltransferase activity in rabbit blood cells obtained from buffy coat layer. *Japan J. Pharmacol.* **55**(Suppl. I): 194P.

31. Kajiyama, K., K. Fujimoto, T. Suzuki, *et al.* 1992. Localization of acetylcholine and choline acetyltransferase activity in human mononuclear leukocytes. *Japan J. Pharmacol.* **58**(Suppl. I): 59P.

32. Rinner, I., K. Kawashima & K. Schauenstein. 1998. Rat lymphocytes produce and secrete acetylcholine in dependence of differentiation and activation. *J. Neuroimmunol.* **81:** 31–37.

33. Hagforsen, E., A. Einarsson, F. Aronsson, *et al.* 2000. The distribution of choline acetyltransferase- and acetylcholinesterase-like immunoreactivity in the palmar skin of patients with palmoplantar pustulosis. *Br. J. Dermatol.* **142:** 234–242.

34. Wessler, I. & C.J. Kirkpatrick. 2008. Acetylcholine beyond neurons: the non-neuronal cholinergic system in humans. *Br. J. Pharmacol.* **154:** 1558–1571.

35. Kawashima, K., N. Watanabe, H. Oohata, *et al.* 1990. Synthesis and release of acetylcholine by cultured bovine arterial endothelial cells. *Neurosci. Lett.* **119:** 156–158.

36. Ikeda, C., I. Morita, A. Mori, *et al.* 1994. Phorbol ester stimulates acetylcholine synthesis in cultured endothelial cells isolated from porcine cerebral microvessels. *Brain Res.* **655:** 147–152.

37. Kirkpatrick, C.J., F. Bittinger, R.E. Unger, *et al.* 2001. The non-neuronal cholinergic system in the endothelium: evidence and possible pathological significance. *Japan. J. Pharmacol.* **85:** 24–28.

38. Haberberger, R.V., M. Bodenbenner, & W. Kummer. 2000. Expression of the cholinergic gene locus in pulmonary arterial endothelial cells. *Histochem. Cell Biol.* **113:** 379–387.

39. Fujii, T., Y. Takada-Takatori & K. Kawashima. 2012. Regulatory mechanisms of acetylcholine synthesis and release by T cells. *Life Sci.* In press.

40. Fujii, T., T. Tsuchiya, S. Yamada, *et al.* 1996. Localization and synthesis of acetylcholine in human leukemic T-cell lines. *J. Neurosci. Res.* **44:** 66–72.

41. Fujii, T., S. Yamada, Y. Watanabe, *et al.* 1998. Induction of choline acetyltransferase mRNA in human mononuclear leukocytes stimulated by phytohemagglutinin, a T-cell activator. *J. Neuroimmunol.* **82:** 101–107.

42. Suenaga, A., T. Fujii, T. Maruyama, *et al.* 2004. Up-regulation of lymphocytic cholinergic activity by ONO-4819, a specific prostaglandin EP4 receptor agonist, in MOLT-3 human leukemic T cells. *Vasc. Pharmacol.* **41:** 51–58.

43. Fujii, T. & K. Kawashima. 2002. Effects of human antithymocyte globulin on acetylcholine synthesis, its release and

choline acetyltransferase transcription in a human leukemic T-cell line. *J. Neuroimmunol.* **128:** 1–8.

44. Fujii, T., K. Masuyama & K. Kawashima. 2006. Simvastatin regulates non-neuronal cholinergic activity in T lymphocytes via CD11a-mediated pathways. *J. Neuroimmunol.* **179:** 101–107.

45. Fujii, T., Y. Watanabe, T. Inoue & K. Kawashima. 2003. Up-regulation of mRNA encoding the M_5 muscarinic acetylcholine receptor in human T- and B-lymphocytes during immunological responses. *Neurochem. Res.* **28:** 423–429.

46. Fujii, T., Y. Takada-Takatori, K. Horiguchi & K. Kawashima. 2012. Mediatophore regulates acetylcholine release from T cells. *J. Neuroimmunol.* **244:** 16–22.

47. Dunant, Y., J.M. Cordeiro & P.P. Goncalves. 2009. Exocytosis, mediatophore, and vesicular Ca^{2+}/H^+ antiport in rapid neurotransmission. *Ann. N.Y. Acad. Sci.* **1152:** 100–112.

48. Bonner, T.I., N.J. Buckley, A.C Young & M.R. Brann. 1987. Identification of a family of muscarinic acetylcholine receptor genes. *Science* **237:** 527–532.

49. Alexander, S.P.H. & J.A. Peters. 1999. Tips receptor and ion channel nomenclature (10th ed.). *Trends Pharmacol. Sci.* **20**(Suppl.): 6–8.

50. Hulme, E.C., N.J.M. Birdsall & N.J. Buckley. 1990. Muscarinic receptor subtypes. *Ann. Rev. Pharmacol. Toxicol.* **30:** 633–673.

51. Sato, K.Z., T. Fujii, Y. Watanabe, *et al.* 1999. Diversity of mRNA expression for muscarinic acetylcholine receptor subtypes and neuronal nicotinic acetylcholine receptor subunits in human mononuclear leukocytes and leukemic cell lines. *Neurosci. Lett.* **266:** 17–20.

52. Abe, K., T. Fujii, S. Iho, *et al.* 2004. Detection of α9 and α10 nicotinic acetylcholine receptor subunits in human T and B lymphocytes. *J. Pharmacol. Sci.* **94** (Suppl. I): 202P.

53. Qian, J., V. Galitovskiy, A.I. Chernyavsky, *et al.* 2011. Plasticity of the murine spleen T-cell cholinergic receptors and their role in in vitro differentiation of naïve CD4 T cells toward the Th1, Th2 and Th17 lineages. *Genes Immun.* **12:** 222–230.

54. Fujii, T. & K. Kawashima. 2000. Calcium oscillation is induced by muscarinic acetylcholine receptor stimulation in human leukemic T- and B-cell lines. *Naunyn-Schmiedberg's Arch. Pharmacol.* **362:** 14–21.

55. Fujii, T. & K. Kawashima. 2000. Calcium signaling and c-fos gene expression via M_3 muscarinic acetylcholine receptors in human T- and B-cells. *Japan. J. Pharmacol.* **84:** 124–132.

56. Fujino, H., Y. Kitamura, T. Yada, *et al.* 1997. Stimulatory roles of muscarinic acetylcholine receptors in T cell antigen receptor/CD3 complex-mediated interleukin-2 production in human peripheral blood lymphocytes. *Mol. Pharmacol.* **51:** 1007–1014.

57. Elgoyhen, A.B., D.E. Vetter, E. Katz, *et al.* 2001. α10: a determinant of nicotinic cholinergic receptor function in mammalian vestibular and cochlear mechanosensory hair cells. *Proc. Natl. Acad. Sci. USA.* **98:** 3501–3506.

58. Kimura, R., N. Ushiyama, T. Fujii & K. Kawashima. 2003. Nicotine-induced Ca^{2+} signaling and down –regulation of

nicotinic acetylcholine receptor subunit expression in the CEM human leukemic T-cell line. *Life Sci.* **72:** 2155–2159.

59. Zimring, J.C., L.M. Kapp, M. Yamada, *et al.* 2005. Regulation of CD8+ cytolytic T lymphocyte differentiation by a cholinergic pathway. *J. Neuroimmunol.* **164:** 66–75.

60. Vezys, V., D. Masioust, M. Desmarets, *et al.* 2007. Analysis of CD8+ T cell-mediated anti-viral responses in mice with targeted deletions of the M1 or M5 muscarinic cholinergic receptors. *Life Sci.* **80:** 2330–2333.

61. Sinha, K., K.D. Dannelly & S.K. Ghosh. 2001. Effects of T-lymphocyte-dependent and -independent immunity on cholinergic enzyme activity in mouse lacrimal gland. *Exp. Physiol.* **86:** 169–176.

62. Fujii, T., Y. Watanabe, K. Fujimoto & K. Kawashima. 2005. Expression of acetylcholine in lymphocytes and modulation of an independent lymphocytic cholinergic activity by immunological stimulation. In *Advances in Neuroregulation and Neuroprotection.* C. Collin, M. Minami, H. Parves, H. Saito, G.A. Qureshi & C. Reiss, Eds.: 737–750. VSP. Leiden and Boston.

63. Fujii, Y.X., A. Tashiro, K. Arimoto, *et al.* 2007. Diminished antigen-specific IgG$_1$ and interleukin-6 production and acetylcholinesterase expression in combined M$_1$ and M$_5$ muscarinic acetylcholine receptor knockout mice. *J. Neuroimmunol.* **188:** 80–85.

64. Vira, A., M.E. Selkirk, M. Darby, *et al.* 2011. Cholinergic signaling via the M3 muscarinic acetylcholine receptor on CD4+ T cells is required for optimal Th2 immunity to parasitic nematode infection. In *The 3rd International Symposium on Non-Neuronal Acetylcholine,* August 24–26, 2011. Groningen, The Netherlands. Abstract book 32P.

65. Wang, H., M. Yu, M. Ochani, *et al.* 2003. Nicotinic acetylcholine receptor α7 subunit is an essential regulator of inflammation. *Nature* **421:** 384–388.

66. Fujii, Y.X., H. Fujigaya, Y. Moriwaki, *et al.* 2007. Enhanced serum antigen-specific IgG$_1$ and proinflammatory cytokine production in nicotinic acetylcholine receptor α7 subunit gene knockout mice. *J. Neuroimmunol.* **189:** 69–74.

67. Skok, M.V., R. Grailhe, F. Agenes & J.P. Changeux. 2007. The roles of nicotinic receptors in B-lymphocyte development and activation. *Life Sci.* **80:** 2334–2336.

68. Adermann, K., F. Wattler, S. Wattler, *et al.* 1999. Structural and phylogenetic characterization of human SLURP-1, the first secreted mammalian member of the Ly-6/uPAR protein superfamily. *Protein Sci.* **8:** 810–819.

69. Chimienti, F., R.C. Hogg, L. Plantard, *et al.* 2003. Identification of SLURP-1 as an epidermal neuromodulator explains the clinical phenotype of Mal de Meleda. *Hum. Mol. Genet.* **12:** 3017–3024.

70. Arredondo, J., A.I. Chernyauski, D.L. Jolkowky, *et al.* 2005. Biological effects of SLURP-1 on human keratinocyte. *J. Invest. Dermatol.* **125:** 1326–1341.

71. Fischer, J., B. Bouadjar, R. Heilig, *et al.* 2001. Mutations in the gene encoding SLURP-1 in Mal de Meleda. *Hum. Mol. Genet.* **10:** 875–880.

72. Moriwaki, Y., K. Yoshikawa, H. Fukuda, *et al.* 2007. Immune system expression of SLURP-1 and SLURP-2, two endogenous nicotinic acetylcholine receptor ligands. *Life Sci.* **80:** 2365–2368.

73. Mastrangeli, R., S. Donini, C.A. Kelton, *et al.* 2003. ARS component B: structural characterization, tissue expression and regulation of the gene and protein (SLURP-1) associated with Mal de Meleda. *Eur. J. Dermatol.* **13:** 560–570.

74. Pettersson, A., S. Nordlander, G. Nylund, *et al.* 2008. Expression of the endogenous, nicotinic acetylcholine receptor ligand, SLURP-1, in human colon cancer. *Auton. Autacoid Pharmacol.* **28:** 109–116.

75. Horiguchi, K., N. Yamashita, S. Horiguchi, *et al.* 2009. Expression of SLURP-1, an endogenous α7 nicotinic acetylcholine receptor allosteric ligand, in murine bronchial epithelial cells. *J. Neurosci. Res.* **87:** 2740–2747.

76. Tjiu, J-W., P-J. Lin, W-H. Wu, *et al.* 2011. SLURP-1 mutation-impaired T-cell activation in a family with mal de Meleda. *Br. J. Dermatol.* **164:** 47–53.

77. Gamse, R., P. Holzer & F. Lembeck. 1980. Decrease of substance P in primary afferent neurons and impairment of neurogenic plasma extravasation by capsaicin. *Br. J. Pharmacol.* **68:** 207–213.

78. Li, Y., Y.C. Jiang & C. Owyang. 1998. Central CGRP inhibits pancreatic enzyme secretion by modulation of vagal parasympathetic outflow. *Am. J. Physiol.* **275:** G957–G963.

79. Moriwaki, Y., Y. Watanabe, T. Shinagawa, *et al.* 2009. Primary sensory neuronal expression of SLURP-1, an endogenous nicotinic acetylcholine receptor ligand. *Neurosci. Res.* **64:** 403–412.

80. Favre, B., L. Plantard, L. Aeschbach, *et al.* 2007. SLURP-1 is a late marker of epidermal differentiation and is absent in Mal de Meleda. *J. Invest. Dermatol.* **127:** 301–308.

Ann. N.Y. Acad. Sci. ISSN 0077-8923

ANNALS OF THE NEW YORK ACADEMY OF SCIENCES

Issue: *Neuroimmunomodulation in Health and Disease*

T cells affect central and peripheral noradrenergic mechanisms and neurotrophin concentration in the spleen and hypothalamus

Jamela Jouda,[1] Johannes Wildmann,[1] Martin Schäfer,[2] Eduardo Roggero,[3] Hugo O. Besedovsky,[1] and Adriana del Rey[1]

[1]Department of Immunophysiology, Institute of Physiology and Pathophysiology, Medical Faculty, Marburg, Germany. [2]Department of Molecular Neuroscience, Institute of Anatomy and Cell Biology, Medical Faculty, Marburg, Germany. [3]Department of Physiology, Faculty of Medicine, Universidad Abierta Interamericana, Rosario, Argentina

Address for correspondence: Adriana del Rey, Department of Immunophysiology, Institute of Physiology and Pathophysiology, Deutschhausstrasse 2, 35037 Marburg, Germany. delrey@mailer.uni-marburg.de

Interactions between T cells and noradrenergic pathways were investigated using athymic nude mice as a model. Higher noradrenaline (NA) concentrations and increased density of noradrenergic fibers were found in the spleen and hypothalamus, but not in the kidney, of 21-day-old *Foxn1ⁿ* (athymic) mice, compared with *Foxn1ⁿ/Foxn1⁺* (heterozygous) littermates. Although no differences in nerve growth factor concentrations were detected, significantly higher brain-derived neurotrophic factor concentrations were found in the spleen and hypothalamus of athymic mice compared with the controls. All of these alterations were abrogated in *Foxn1ⁿ* mice reconstituted by thymus transplantation at birth. These results suggest that T lymphocytes or their products can induce (1) a decrease in the number and activity in splenic sympathetic nerve fibers; (2) a decrease in NA content in the hypothalamus, which, in turn, may influence the pituitary–adrenal axis and the descending neural pathways associated with the autonomic nervous system; and (3) changes in neurotrophin concentration in the spleen and hypothalamus.

Keywords: athymic mice; T cells; BDNF; NGF; spleen; hypothalamus

Introduction

In several mammalian species, sympathetic noradrenergic nerve fibers innervate lymphoid organs, such as the spleen, thymus, and lymph nodes.[1–3] During ontogeny, the sympathetic innervation of the spleen develops in parallel to the appearance of mature T cells in this organ.[4–6] In adult animals, sympathetic fibers in the spleen follow the central arteriole and penetrate the periarteriolar sheath, a region where mainly T lymphocytes are located. This particular distribution results in close contact between sympathetic fibers and T cells.[2]

Noradrenergic neuronal fibers in the hypothalamus, which mainly arise from cell bodies in brain stem nuclei, modulate the activity of efferent pathways to the pituitary and to the descending brainstem and spinal cord regions associated with the autonomous nervous system.[7,8] In general, the activation of the hypothalamus–pituitary–adrenal (HPA) axis inhibits inflammatory mechanisms and regulates both the extent and the specificity of an ongoing immune response.[9] On the other hand, immune cells release products that have the capacity to affect central and peripheral noradrenergic mechanisms[10,11] and to induce endocrine changes[12,13] that are relevant for immunoregulation.[9,14]

Neurotrophins, a family of structurally related proteins, are involved in the development of nerve fibers both at peripheral and central levels by controlling neuronal survival, development, function, and plasticity.[15–18] Nerve growth factor (NGF) and brain-derived neurotrophic factor (BDNF) are among these neurotrophins. There is evidence that the levels of NGF in several tissues that are targets of sympathetic neurons correlate with the density of sympathetic innervation,[19] and that these tissues are the major source of NGF required by sympathetic neuron survival and functioning.[20] Although

doi: 10.1111/j.1749-6632.2012.06642.x

Ann. N.Y. Acad. Sci. 1261 (2012) 18–25 © 2012 New York Academy of Sciences.

originally isolated from the submaxillary gland, NGF is synthesized by several cell types, including smooth muscle cells, fibroblasts, and neurons, and it was the first neurotrophin shown to be expressed by immune cells (T and B lymphocytes, macrophages, and mast cells).[21–23] BDNF, which was previously thought to be found primarily in neurons in the central nervous system, is also produced by muscle cells and by developing and mature sympathetic neurons.[24] More recently, it has been shown that BDNF is also expressed in immune cells,[21–23] and that it can be produced *in vitro* by all major immune cell types, including CD4[+] and CD8[+] T lymphocytes, B lymphocytes, and monocytes.[23]

There is good evidence indicating that the nervous, endocrine, and immune systems interact with each other, and it is expected that these interactions are disturbed when the functioning of one system is altered. Congenitally athymic nude mice offer an excellent model to study the relevance of T lymphocytes in these interactions.[25] For example, we have previously shown that the absence of T cells in con-

genitally athymic mice results in changes in splenic sympathetic innervation, as reflected by increased noradrenaline (NA) concentrations.[26] In this work, we used athymic (*Foxn1[n]*) mice, normal thymus-bearing heterozygous littermates (*Foxn1[n]/Foxn1[+]*), and athymic nude mice reconstituted by thymus implantation at birth to study the effect of T cells on noradrenergic pathways in the brain and in the sympathetic nervous system (SNS). We also determined the concentration of the neurotrophins NGF and BDNF in the spleen and hypothalamus of the same animals.

Materials and methods

Animals

Male athymic (Balb/c *Foxn1[n]*), heterozygous thymus-bearing littermates (Balb/c *Foxn1[n]/Foxn1[+]*) and athymic mice that were reconstituted with thymus at birth were used for these studies. Mice were bred under conventional conditions, in temperature- and light- (12-h light–dark cycles) controlled rooms.

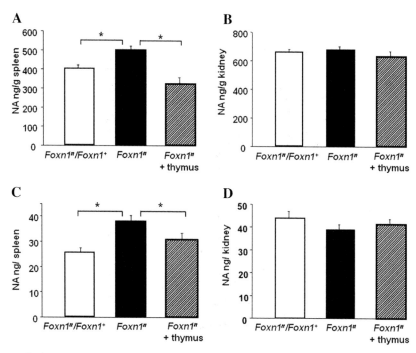

Figure 1. Increased splenic NA concentration in athymic nude mice is normalized by thymus implantation at birth. NA concentration and content in the spleen (A, C) and kidney (B, D) of heterozygous, thymus-bearing (*Foxn1[n]/Foxn1[+]*), athymic (*Foxn1[n]*), and athymic male mice grafted with thymus at birth (*Foxn1[n]* + thymus) were determined when animals were 21 days old. Results are shown as mean ± SD of determinations performed in 10 mice per group. The horizontal bars over the columns indicate statistically significant differences ($P < 0.05$) between the groups.

Thymus transplantation

Two thymi from newborn Balb/c *Foxn1ⁿ/Foxn1⁺* donors were grafted into the axillary region of athymic nude Balb/c *Foxn1ⁿ* mice when they were less than 24 h old.

Organ collection

Groups of mice were sacrificed by cervical dislocation when they were 21 days old. Blood, spleen, kidneys (right and left), and the hypothalamus were collected. Each spleen was cut into three pieces and the hypothalamus was divided into left and right parts. One piece of the spleen, one kidney, and one half of the hypothalamus were fixed in Bouin–Holland for immunhistochemistry. The remaining parts were immediately frozen and kept at −80 °C until used for neurotransmitter and neurotrophin determinations.

NA determination

NA determinations were performed in one-third of the spleen, the left kidney, and the left half of the hypothalamus by high-pressure liquid chromatography (HPLC) as previously described.[27]

Neurotrophin determinations

The middle-third part of the spleen, the right kidney, and the right part of the hypothalamus were used for neurotrophin determinations. A protease inhibitor mixture containing 100 mM 6-amino carpoic acid, 10 mM EDTA, 5 mM benzamidine-HCL, 0.2 mM phenylmethylsulfonyl fluoride was added to the samples and the tissue was disrupted by ultrasonication. Sonicated samples were centrifuged at $20{,}000 \times g$ at 4 °C for 10 min, and the supernatants were collected and stored at −80 °C until used for neurotrophin determinations. NGF and BDNF concentrations in the samples were determined by commercially available enzyme-linked immunosorbent assays (ELISAs; Human β-NGF and β-BDNF Duo set from R&D Development System, Minneapolis, MN). Total protein concentration in the supernatants was determined by Bradford protein assays (Pierce, Rockford, IL).

Corticosterone determination

Corticosterone concentrations in plasma were determined by a solid phase, competitive binding (ELISA, IBL, Hamburg, Germany).

Immunohistochemistry

An indirect immunofluorescent staining method was used to detect tyrosine hydroxylase, the rate-limiting enzyme of NA synthesis and a marker of noradrenergic neurons. After fixing in Bouin–Holland for 48 h and dehydration in a graded series of 2-propanol solutions, the tissues were embedded in Paraplast plus (Merck, Darmstadt, Germany). Adjacent sections (7-μm thick) were cut and deparaffinized. To increase the sensitivity of detection, antigen retrieval was performed by heating the sections at 92–95 °C for 15 min in 0.01 M citrate buffer (pH 6). Nonspecific binding sites were blocked with 5% bovine serum albumin (BSA, Serva, Heidelberg, Germany) in phosphate buffered saline (PBS) followed by an avidin-biotin blocking step (Avidin-biotin Blocking kit, Boehringer Ingelheim, Germany). Sections were incubated overnight at 4 °C with a polyclonal sheep antityrosine hydroxylase affinity purified antibody (Chemicon, Temecula, CA) at a final dilution of 1:200 and further incubated for 2 h at 37 °C. After washing in distilled water and in 50 mM PBS, the sections were incubated with a biotinylated secondary antibody against sheep immunoglobulin (Dianova, Hamburg, Germany) for 45 min at 37 °C, washed several times, and incubated for 2 h at 37 °C with Biotin-SP–conjugated AffiniPure donkey anti-sheep IgG (Dianova,

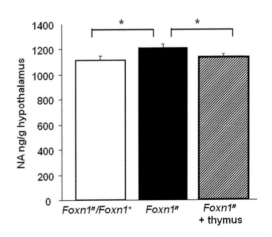

Figure 2. Increased NA concentration in the hypothalamus of athymic nude mice is normalized by thymus implantation at birth. NA concentration was determined in the hypothalamus of the same mice from which the results shown in Figure 1 were obtained. Results are shown as mean ± SD of determinations performed in 10 mice per group. The horizontal bars over the columns indicate statistically significant differences ($P < 0.05$) among the groups.

Hamburg, Germany). Immunoreactions were visualized with streptavidin-Alexa Fluor 488 conjugate. Sections were analyzed and photographed using an Olympus Fluoview Laser Scanning Microscope (Olympus Optical, Hamburg, Germany).

Statistical analysis

Results are expressed as mean ± SE. Data were analyzed by one-way analysis of variance (ANOVA) followed by Fisher's test for multiple comparisons, using StatView version 5.0. Differences were considered significant when $P < 0.05$.

Results

Higher splenic NA levels were detected in the athymic nude $Foxn1^n$ mice than in the heterozygous $Foxn1^n/Foxn1^+$ thymus-bearing littermates. This alteration was completely normalized when the $Foxn1^n$ mice were implanted with a thymus at birth (Fig. 1A). To assert whether the increased NA levels in athymic mice were restricted to the spleen or whether they reflected a general hyperactivity of the SNS, NA levels of another abdominal organ, the kidney, were determined. No differences among the NA levels in the kidneys of athymic-$Foxn1^n$, $Foxn1^n/Foxn1^+$, or $Foxn1^n$ mice implanted

with thymus at birth were detected (Fig. 1B). Because $Foxn1^n$ mice have smaller spleens and kidneys do than $Foxn1^n/Foxn1^+$ animals, the results were also expressed as total NA content of the organs studied. The total NA content in the spleen of $Foxn1^n$ animals was also significantly higher than that of $Foxn1^n/Foxn1^+$ mice (Fig. 1C), but no significant differences were found in the kidneys (Fig. 1D). Higher NA levels in the hypothalamus were observed in 21-day-old $Foxn1^n$ mice compared with $Foxn1^n/Foxn1^+$ mice, and the NA concentration was completely normalized when $Foxn1^n$ mice were implanted with a thymus at birth (Fig. 2).

These results prompted us to perform immunohistochemical studies in the spleen and hypothalamus of $Foxn1^n/Foxn1^+$ mice, athymic $Foxn1^n$ mice, and $Foxn1^n$ mice implanted with a thymus at birth. These studies confirmed the results of the NA determinations, namely, there were clear differences among the various groups both in the number and intensity of tyrosine hydroxylase–positive fibers in the spleen and hypothalamus. Athymic $Foxn1^n$ mice showed enhanced fluorescence intensity and the number of fluorescent fibers was higher compared with the heterozygous $Foxn1^n/Foxn1^+$ mice and with the thymus-grafted $Foxn1^n$ mice

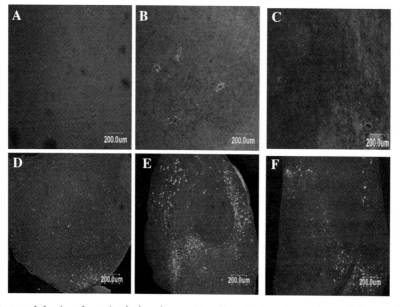

Figure 3. The increased density of tyrosine hydroxylase–positive fibers in the spleen and hypothalamus of athymic mice is normalized by neonatal thymus graft. The photos show representative examples of tyrosine hydroxylase–containing fibers in the spleen (A, B, C) and hypothalamus (D, E, F) of 21-day-old thymus-bearing $Foxn1^n/Foxn1^+$ (A, D), athymic nude $Foxn1^n$ mice (B, E), and $Foxn1^n$ mice to which a thymus was transplanted at birth (C, F).

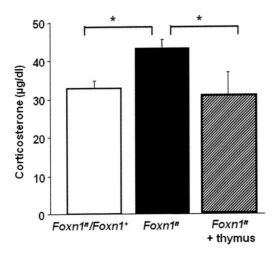

Figure 4. Increased corticosterone blood levels in athymic nude mice are normalized by thymus implantation at birth. Corticosterone plasma levels were determined in the same mice from which the results shown in Figures 1 and 2 were obtained. Results are shown as mean ± SD of determinations performed in 10 mice per group. The horizontal bars over the columns indicate statistically significant differences ($P < 0.05$) among the groups.

(Fig. 3A–C). These differences were even more marked in the hypothalamus (Fig. 3D–F). Corticosterone blood levels were higher in 21-day-old *Foxn1^n* mice than in *Foxn1^n/Foxn1^+* mice and the levels normalized in thymus-implanted *Foxn1^n* mice (Fig. 4). NGF and BDNF concentrations in spleen, kidney, and hypothalamus were also determined (Fig. 5). No statistically significant differences between the three groups were observed in NGF concentration in spleens and in the hypothalamus (Figs. 5A and B). However, BDNF concentrations in the spleen and hypothalamus, but not in the kidney, were higher in athymic *Foxn1^n* mice compared with the thymus-bearing *Foxn1^n/Foxn1^+* littermates. Thymus graft at birth to *Foxn1^n* mice resulted in BDNF concentrations that were comparable to those of the *Foxn1^n/Foxn1^+* littermates (Figs. 5C and D).

Discussion

The present results show that athymic nude *Foxn1^n* mice have higher splenic NA content and concentration and more noradrenergic fibers compared with their normal thymus-bearing *Foxn1^n/Foxn1^+* littermates. These alterations do not reflect a general sympathetic hyperactivity in nude mice because

no comparable differences are detected in a non-lymphoid organ, such as the kidney. The fact that neonatal thymus grafting to newborn nude mice normalizes splenic NA content and the number of noradrenergic fibers indicates that these alterations are causally related to the absence of the thymus. It is known that the sympathetic innervation of several peripheral organs in rodents is not completely developed until relatively late in ontogeny, that is, weeks after birth. This includes the spleen,[5] in which the T cell compartment develops late in postnatal life.[6] In the particular case of athymic nude mice, such a T cell compartment is not present because T cell lymphopoiesis is interfered with early in embryogenesis,[28] and the thymus-dependent areas in the spleen of nude mice are not populated by mature T cells.

We also report here that the hypothalamus of athymic *Foxn1^n* mice has a higher NA concentration and number of noradrenergic fibers compared with of their normal thymus-bearing littermates, and that thymus grafting into newborn nude mice results in normalization of these parameters, indicating that these alterations are caused by the absence of the thymus. This interpretation is in line with earlier studies showing that lymphokine-containing supernatants from T cells have the capacity to evoke changes in central noradrenergic neurons.[10]

We also show here that athymic nude *Foxn1^n* mice have higher corticosterone blood levels than their normal thymus-bearing littermates, and that neonatal thymus grafting into nude mice normalizes the levels of this hormone. Our results are in agreement with the evidence that an increase in hypothalamic NA concentration results in stimulation of the HPA axis.[29] Furthermore, they indicate that T cells or their products can affect noradrenergic pathways in the hypothalamus that might lead to differences in the functioning of the HPA axis.

The fact that the alterations in central and peripheral noradrenergic systems observed in athymic mice can be reversed by thymus implantation at birth allows the conclusion that such alterations are not just a genetic epiphenomenon fortuitously associated with the absence of mature T cells. Because our previous studies in nude mice showed that the alteration in peripheral noradrenergic pathways can also be reversed by T cell inoculation,[26] it is unlikely that thymic hormones are causally related to the effects reported here. Rather, it seems more likely that

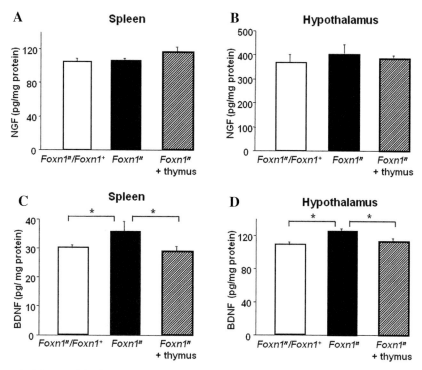

Figure 5. Congenital absence of the thymus results in increased BDNF concentration in the spleen and hypothalamus: reversal by neonatal thymus graft. NGF (A, B) and BDNF (C, D) concentrations were determined in the spleen and hypothalamus of 21-day-old heterozygous, thymus-bearing (*Foxn1^n/Foxn1$^+$*), athymic (*Foxn1^n*), and athymic mice grafted with thymus at birth (*Foxn1^n* + thymus). Results are expressed as pg neurotrophin per mg protein in the corresponding tissue, and shown as mean ± SD (NGF: $n = 10$ per group; BDNF: $n = 7$–9 per group). No statistically significant differences in NGF concentrations were detected between the groups. The horizontal bars over the columns indicate statistically significant differences in BDNF concentrations among the groups ($P < 0.05$).

T cells, acting either directly or indirectly, exert an inhibitory influence on the development of splenic sympathetic innervation and in the noradrenergic pathways that project to the hypothalamus.

Neurotrophins are essential for neuronal growth and differentiation during development, but they are also important regulators of survival and maintenance of nerve cells during adulthood. Thus, as an attempt to understand the mechanisms that may cause increased activity in central and peripheral noradrenergic nerve fibers, we determined the concentration of two neurotrophins in parallel to the levels of NA. No significant differences were detected in NGF concentrations, but athymic nude mice had higher BDNF levels in the spleen and hypothalamus than did their normal thymus-bearing littermates. This was not a general occurrence because no significant differences between athymic and euthymic

mice were found in another organ, such as the kidney. Interestingly, besides normalizing NA levels, thymus grafting at birth also normalized BDNF concentrations in the spleen and in the hypothalamus. Taken together, these results allow the conclusion that higher BDNF, but not NGF, concentrations are associated with the absence of mature T cells.

In summary, the results described indicate that the known effects of sympathetic neurotransmitters and of neuroendocrine mechanisms under the control of NA brain neurons on immune cell activity do not represent a unidirectional process, but that such influences are bidirectional. Indeed, T cells can also influence the development of neural structures. The present results in nude mice are in agreement with the view that, in general, immune cells play an inhibitory role on the development of the sympathetic innervation of lymphoid organs and/or in the activity of the SNS and of central NA neurons. Indeed, we have also shown that reduced immunological activity in germ-free animals results in increased basal NA levels in lymphoid organs,[30] and that NA levels in the spleen and NA turnover rates in the hypothalamus decrease during the course of specific immune responses.[31,32] More recently, we showed that the sympathetic innervation of the spleen is markedly reduced in mice with a hyperactive immune system, such as *lpr/lpr* mice, which develop a lymphoproliferative/autoimmune disease. These mutual immune-neural effects reinforces the concept that a neuroendocrine-immune network of interactions operates during development and adult life.

Conflicts of interest

The authors declare no conflicts of interest.

References

1. Reilly, F.D., R.S. McCuskey & H.A. Meineke. 1976. Studies of the hemopoietic microenvironment. VIII. Andrenergic and cholinergic innervation of the murine spleen. *Anat. Rec.* **185:** 109–117.

2. Williams, J.M. & D.L. Felten. 1981. Sympathetic innervation of murine thymus and spleen: a comparative histofluorescence study. *Anat. Rec.* **199:** 531–542.

3. Bulloch, K. 1985. Neuroanatomy of lymphoid tissue: a review. In *Neural Modulation of Immunity*. R. Guillemin, M. Cohn, & T. Melnechuk, Eds.: 111–141. Raven Press. New York.

4. Wu, S., F.H. Bach & R. Auerbach. 1975. Cell-mediated immunity: differential maturation of mixed leukocyte reaction and cell-mediated lympholysis. *J. Exp. Med.* **142:** 1301–1305.

5. De Champlain, J. & B. Smith. 1974. Ontogenesis of central and peripheral adrenergic neurons in the rat following neonatal treatment with 6-hydroxydopamine. In *Dynamics of Degeneration and Growth in Neurons*. K. Fuxe, L. Olsen & Y. Zotterman, Eds.: 435–445. Pergamon Press. Oxford, New York.

6. Bach, J., *et al.* 1975. T cell subsets: terminology problems. In *The Biochemical Activity of Thymic Hormones*. D.W. van Bekkum, Ed.: 159–168. Wiley. New York.

7. Livnat, S., *et al.* 1985. Involvement of peripheral and central catecholamine systems in neural-immune interactions. *J. Neuroimmun.* **10:** 5–30.

8. Carlson, S.L., *et al.* 1987. Alterations of monoamines in specific central autonomic nuclei following immunization in mice. *Brain Behav. Immun.* **1:** 52–63.

9. Sapolsky, R.M., L.M. Romero & A.U. Munck. 2000. How do glucocorticoids influence stress responses? Integrating permissive, suppressive, stimulatory, and preparative actions. *Endocr. Rev.* **21:** 55–89.

10. Besedovsky, H., *et al.* 1983. The immune response evokes changes in brain noradrenergic neurons. *Science* **221:** 564–566.

11. Kabiersch, A., *et al.* 1988. Interleukin-1 induces changes in norepinephrine metabolism in the rat brain. *Brain Behav. Immun.* **2:** 267–274.

12. Besedovsky, H., *et al.* 1986. Immunoregulatory feedback between interleukin-1 and glucocorticoid hormones. *Science* **233:** 652–654.

13. Malarkey, W.B. & P.J. Mills. 2007. Endocrinology: the active partner in PNI research. *Brain Behav. Immun.* **21:** 161–168.

14. Elenkov, I.J., *et al.* 2000. The sympathetic nerve–an integrative interface between two supersystems: the brain and the immune system. *Pharmacol. Rev.* **52:** 595–638.

15. Lewin, G.R. & Y.A. Barde. 1996. Physiology of the neurotrophins. *Annu. Rev. Neurosci.* **19:** 289–317.

16. Katoh-Semba, R., *et al.* 1989. Sex-dependent and sex-independent distribution of the beta-subunit of nerve growth factor in the central nervous and peripheral tissues of mice. *J. Neurochem.* **52:** 1559–1565.

17. Lommatzsch, M., *et al.* 1999. Abundant production of brain-derived neurotrophic factor by adult visceral epithelia. Implications for paracrine and target-derived Neurotrophic functions. *Am. J. Pathol.* **155:** 1183–1193.

18. Helke, C.J., *et al.* 1998. Axonal transport of neurotrophins by visceral afferent and efferent neurons of the vagus nerve of the rat. *J. Comp. Neurol.* **393:** 102–117.

19. Korsching, S. & H. Thoenen. 1983. Quantitative demonstration of the retrograde axonal transport of endogenous nerve growth factor. *Neurosci. Lett.* **39:** 1–4.

20. Korsching, S. & H. Thoenen. 1985. Treatment with 6-hydroxydopamine and colchicine decreases nerve growth factor levels in sympathetic ganglia and increases them in the corresponding target tissues. *J. Neurosci.* **5:** 1058–1061.

21. Besser, M. & R. Wank. 1999. Cutting edge: clonally restricted production of the neurotrophins brain-derived neurotrophic factor and neurotrophin-3 mRNA by human immune cells and Th1/Th2-polarized expression of their receptors. *J. Immunol.* **162:** 6303–6306.

22. Kerschensteiner, M., *et al.* 1999. Activated human T cells, B cells, and monocytes produce brain-derived neurotrophic factor in vitro and in inflammatory brain lesions: a neuroprotective role of inflammation? *J. Exp. Med.* **189:** 865–870.

23. Kerschensteiner, M., *et al.* 2003. Neurotrophic cross-talk between the nervous and immune systems: implications for neurological diseases. *Ann. Neurol.* **53:** 292–304.

24. Causing, C.G., *et al.* 1997. Synaptic innervation density is regulated by neuron-derived BDNF. *Neuron* **18:** 257–267.

25. Loor, F. & B. Kindred. 1974. Immunofluorescence study of T cell differentiation in allogeneic thymus-grafted nudes. In *Proceedings of the First International Workshop on Nude Mice*. J. Rygaard & C. Povsen, Eds.: 141–147. Gustav-Fisher-Verlag. Stuttgart, Germany.

26. Besedovsky, H.O., *et al.* 1987. T lymphocytes affect the development of sympathetic innervation of mouse spleen. *Brain Behav. Immun.* **1:** 185–193.

27. del Rey, A., *et al.* 2008. Disrupted brain-immune system-joint communication during experimental arthritis. *Arthritis Rheum.* **58:** 3090–3099.

28. Pritchard, H. & H.S. Micklem. 1974. The nude (nu/nu) mouse as a model of thymus and T-lymphocyte deficiency. In *Proceedings of the First International Workshop on Nude Mice*. J. Rygaard & C. Povlsen, Eds.: 127–139. Gustav-Fisher-Verlag. Stuttgart, Germany.

29. Itoi, K., *et al*. 1994. Microinjection of norepinephrine into the paraventricular nucleus of the hypothalamus stimulates corticotropin-releasing factor gene expression in conscious rats. *Endocrinology* **135:** 2177–2182.

30. del Rey, A., *et al*. 1981. Immunoregulation mediated by the sympathetic nervous system, II. *Cell Immunol.* **63:** 329–334.

31. Besedovsky, H.O., *et al*. 1979. Immunoregulation mediated by the sympathetic nervous system. *Cell Immunol.* **48:** 346–355.

32. del Rey, A., *et al*. 1982. Sympathetic immunoregulation: difference between high- and low-responder animals. *Am. J. Physiol.* **242:** R30–R33.

Ann. N.Y. Acad. Sci. ISSN 0077-8923

Sympathetic nerve fiber repulsion: testing norepinephrine, dopamine, and 17β-estradiol in a primary murine sympathetic neurite outgrowth assay

Susanne Klatt, Alexander Fassold, and Rainer H. Straub

Laboratory of Experimental Rheumatology and Neuroendocrine Immunology, Division of Rheumatology, Department of Internal Medicine I, University Hospital, Regensburg, Regensburg Germany

Address for correspondence: Rainer H. Straub, Laboratory of Experimental Rheumatology and Neuroendocrine Immunology, Division of Rheumatology, Department of Internal Medicine I, University Hospital, 93042 Regensburg, Germany. rainer.straub@klinik.uni-regensburg.de

Loss of sympathetic nerve fibers (SNFs) occurs in inflamed tissue; and select semaphorins, upregulated during inflammation, stimulate repulsion/loss of SNFs. However, it is unknown whether other factors released locally in inflamed tissue, such as norepinephrine, dopamine, and 17β-estradiol, are also repellent. In order to study the effects of hormones on SNF repulsion, an SNF outgrowth assay was used. The repellent activity of semaphorins 3C was weaker than of semaphorin 3F. Tumor necrosis factor α (TNF-α) repelled nerve fibers with moderate to strong effects (from 0–100% repulsion). High concentrations of dopamine and norepinephrine (10^{-6} M) induced weak but significant nerve fiber repulsion (up to 20%). Norepinephrine at 10^{-8} M was comparable with 10^{-6} M at inducing nerve fiber outgrowth. Stimulation with low concentrations of 17β-estradiol (10^{-10} M, but not 10^{-8} M) repelled SNFs. These results demonstrate that not only specific axon guidance molecules, such as semaphorins 3F and 3C, but also hormonal factors and TNF-α influence SNF repulsion and outgrowth.

Keywords: Sympathetic nervous system; nerve repellent factors; TNF-α; catecholamines; 17β-estradiol; rheumatoid arthritis

Introduction

During the last few years it has become clear that the sympathetic nervous system is involved in the pathogenesis of arthritis. Recent studies have revealed that sympathetic nerve fibers (SNFs) are lost in inflamed synovial tissue.[1] The loss of SNFs was also observed in inflammatory lesions in experimental colitis, Crohn's disease, Charcot's arthropathy, experimental insulitis, and other inflammatory processes,[2–5] indicating that SNF loss is a general principle in inflamed tissue. It is widely accepted that neurotransmitters of the sympathetic nervous system act—at high concentrations—as anti-inflammatory mediators by inhibiting the secretion of tumor necrosis factor α (TNF-α), interleukin-2 (IL-2), IL-12, and other proinflammatory signals relevant in rheumatoid arthritis (RA), osteoarthri-

tis, and experimental arthritis.[6] Norepinephrine and adenosine can bind to different receptors with opposing effects, depending on the local concentration of these sympathetic neurotransmitters.[7] At high concentrations, an anti-inflammatory effect produced by inhibiting TNF-α secretion is mediated via β2-adrenergic and A2 adenosine receptors.[8] At low concentrations, norepinephrine and adenosine bind to α1/2-adrenoreceptors and A1 adenosine receptors, which stimulate proinflammatory signals.[9]

The anti-inflammatory effects of sympathetic neurotransmitters depend on the presence of anti-inflammatory SNFs, which are possibly lost within hours after outbreak of inflammation, as complete repulsion of outgrown neurites can be observed within 5 h *in vitro*.[10] The loss of these nerve fibers and, at the same time, of neurotransmitters is most likely a consequence of the initial

doi: 10.1111/j.1749-6632.2012.06628.x

synovial inflammation. The mechanisms of SNF loss in inflamed tissue are not understood. One possible explanation is repulsion by specific nerve-repellent factors of the semaphorin group.[11,12] Semaphorin 3C and semaphorin 3F as specific repellent factors of SNFs (reviewed in Refs. 13 and 14) are upregulated in synovial fibroblasts and macrophages during the inflammatory process.[10,15] They bind to their surface receptor neuropilin-2 and its coreceptor plexin A2, which are located on the surface of sympathetic nerve terminals.[13,16] Recent studies by our group showed that semaphorin 3F is a nerve repellent factor.[10] In these experiments, however, we did not focus on other factors that might influence repulsion or attraction of SNFs. The repellent activity of semaphorin 3C, for example, was not tested in our nerve outgrowth assay. High levels of proinflammatory TNF-α exist in rheumatoid synovium, but TNF-α has also not been investigated in the context of SNF repulsion in our outgrowth assay. TNF-α, however, may influence the function of SNFs because TNF-α receptors are present on SNFs.[17,18]

Synovial macrophages were identified as a source of norepinephrine and dopamine in synovial tissue.[1] B cells, macrophages, fibroblasts, mast cells, and granulocytes have recently been identified as catecholamine-producing cells,[19–21] and these neurotransmitters are locally produced in the inflamed synovium.[19] The two neurotransmitters are present in the local environment of SNFs in synovial tissue, independent of SNF function. This may influence the behavior of SNFs, which in turn might be an additional factor that influences nerve fiber density.[22] Since sympathetic nerve endings possess catecholamine-responsive receptors, effects of catecholamines are to be expected. However, the importance of norepinephrine and dopamine for SNF repulsion or outgrowth is not known.

Another important neuroendocrine hormone in inflamed tissue is 17β-estradiol. Data indicate that female sex hormones play an important role in the pathophysiology of chronic inflammatory diseases like RA.[23] Interestingly, the concentration of 17β-estradiol is elevated in superfusate samples of synovial tissue. Thus, we hypothesized that the female sex hormone 17β-estradiol might be important for the process of SNF repulsion. Because estrogen receptors exist in sympathetic nerve fibers,[24,25] 17β-estradiol might also affect SNF repulsion or outgrowth.

Because of the critical role of SNFs in arthritis and other inflammatory diseases, the investigation of additional factors that influence repulsion or outgrowth independent of specific repellent factors of the semaphorin group is of importance. In the study presented below, we investigated the effects of TNF-α, norepinephrine, dopamine, and 17β-estradiol in a recently established neurite outgrowth assay.

Methods

Materials

Semaphorin 3F, semaphorin 3C, and TNF-α were purchased from R&D Systems (Wiesbaden, Germany). The hormones and neurotransmitters 17β-estradiol, norepinephrine, and dopamine were purchased from Sigma (Munich, Germany).

Removal of sympathetic ganglia and culture

The removal and culture of sympathetic ganglia was carried out as described previously.[10] All animal procedures were in accordance with the German Animal Welfare Act. Sympathetic ganglia of 2-day-old C57BL/6 mouse pups were dissected under a microscope (Zeiss Stemi 2000 stereo microscope, Zeiss; Jena, Germany). They were digested with dispase (Roche; Mannheim, Germany) to ease outgrowth of SNF and then transferred to poly-D-lysine-coated culture slides (BD Biosciences; Heidelberg, Germany) in Ham's F12 + GlutaMAX medium (Invitrogen; Karlsruhe, Germany). SNF outgrowth was stimulated by 100 ng/mL of nerve growth factor 7S (Sigma; Deisenhofen, Germany). After 2 days in a 5% CO_2 humidified atmosphere at 37 °C, the growth medium was replaced with Krebs-Ringer solution (Sigma) buffered with 20 mmoles HEPES (Sigma) at a pH of 7.4. At this time point, the neurites had a length of approximately three millimeters. In this outgrowth stage, semaphorins 3C and 3F, TNF-α, norepinephrine, dopamine, and 17β-estradiol were added at different concentrations. Experiments were conducted according to institutional and governmental regulations for animal use.

Quantification of outgrown and repelled axons

The visualization of the growth behavior of nerve fibers was performed in an incubator for life cell imaging at 37 °C (Ibidi; Martinsried, Germany). Images of viable ganglia *in vitro* were obtained using

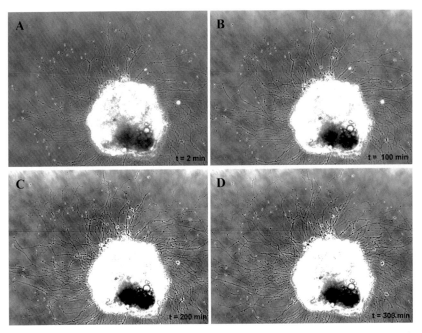

Figure 1. The outgrowth of sympathetic nerve fibers in a neurite outgrowth assay. Depicted are images of stimulated sympathetic nerve fibers from murine sympathetic trunk ganglia of C57BL/6 mice at different time points. (A) At the beginning of the experiment (2 min), 38 protruding axons were marked with a yellow cross (small x). (B) At 100 min. (C) At 200 min. (D) At the end of the experiment (300 min) outgrown axons ($n = 25$) were marked with a green cross and repelled axons with a red cross. The numbers of repelled and further outgrown axons were expressed relative to the total number of axons included in the analysis. Magnification 10×.

a microscope with a life-imaging camera (Zeiss Axiovert 200 MAT microscope; Zeiss AxioCam Mrm videocamera). In order to quantify repulsion or outgrowth, at the beginning of the experiment the main protruding axons were marked with a yellow cross using life images of the ganglion (Zeiss Software AxioVision), which was necessary in order to estimate repulsion or outgrowth in the course of nerve fiber observation.[10] At the end of the experiment (300 min), the numbers of repelled or outgrown axons were counted and expressed relative to the number of all axons included in the analysis (all yellow crosses) (Fig. 1). The digital images were processed with Adobe Photoshop 7.0 (Adobe Systems, San Jose, CA).

Presentation of data and statistical analysis

All data were given as box plots. The group medians were compared by the nonparametric Mann–Whitney test (SigmaPlot, Version 11; Systat, Erkrath, Germany). *P* values less than 0.05 were considered significant.

Results

Effects of semaphorins, TNF-α, and 17β-estradiol in primary murine sympathetic neurite outgrowth assay

As reported recently, we established a neurite outgrowth assay using murine sympathetic ganglia of the thoracic sympathetic trunk from newborn C57BL/6 mice, of which, approximately 25% grew out to build a fine web of SNFs (Fig. 1). Semaphorin 3F induced a dose-dependent repulsion of nerve fibers up to 100% (Fig. 2A). The median repulsion with 10^{-7} M of semaphorin 3F was 80% (Fig. 2A). Concomitantly, semaphorin 3F decreased further outgrowth of SNFs at 10^{-7} M (Fig. 2B).

Because we hadn't previously studied semaphorin 3C, this repellent factor was tested in our neurite outgrowth assay. The effect of semaphorin 3C was markedly smaller compared with semaphorin 3F (Fig. 2C). The median repulsion at a concentration of 10^{-6} M was 45% (Fig. 2C). At the same concentration, further outgrowth was completely inhibited (Fig. 2D). In addition, to achieve the same

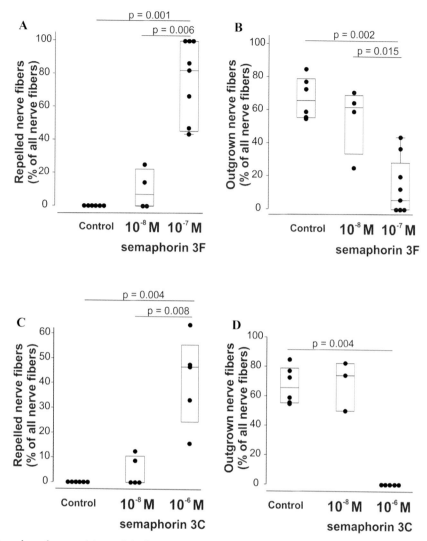

Figure 2. Dose-dependent repulsion and further outgrowth of sympathetic nerve fibers under semaphorins 3F and 3C. (A) Repulsion of sympathetic axons by semaphoring 3F. Without the addition of semaphorin 3F to the medium, no repulsion occurs (control). (B) Outgrowth of sympathetic axons at different doses of semaphorin 3F. Further outgrowth is considered normal without the effects of repellent factors due to the prior addition of nerve growth factor to the medium. (C) Repulsion of sympathetic axons induced by semaphorin 3C. (D) Further outgrowth of sympathetic axons in the presence of different concentrations of semaphorin 3C. Each circular area represents one ganglion (observation of growth behavior over 5 hours).

effect, the concentration of semaphorin 3C needed to be 10 times higher compared with semaphorin 3F (compare Figs. 2A and C, and 2B and D).

Because TNF-α is an abundant cytokine in inflamed tissue, we tested TNF-α for possible SNF repulsion. In some cases, strong repulsion was observed, but sometimes only weak repulsion was measurable (Fig. 3A). Similarly, TNF-α sometimes

inhibited further outgrowth of SNFs, but the median was not significantly different from the control median (Fig. 3B).

Furthermore, 17β-estradiol–induced repulsion of SNFs at both concentrations were tested, but the effect was only significant at 10^{-10} M (Fig. 3C). At a concentration of 10^{-8} M, SNFs grew out more intensely than at 10^{-10} M (Fig. 3D).

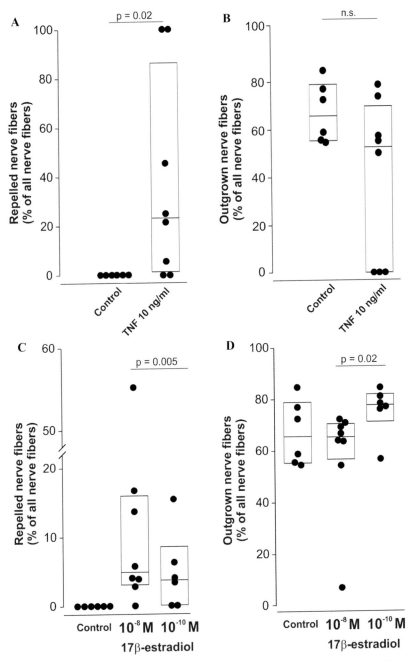

Figure 3. Effects of TNF-α and 17β-estradiol on repulsion and further outgrowth of sympathetic nerve fibers. (A) Repulsion of sympathetic axons at a concentration of 10 ng/mL TNF-α. (B) Outgrowth of axons under TNF-α. (C) Repulsion of sympathetic axons by 17β-estradiol. (D) Outgrowth of axons under 17β-estradiol. Each circular area represents one ganglion (observation of growth behavior over 5 hours). Abbreviations: n.s., not significant.

Effects of catecholamines in primary murine sympathetic neurite outgrowth assay

Norepinephrine and dopamine can be measured in supernatants of mixed synovial cells of RA pa-tients.[19] Thus, both catecholamines were tested for a possible influence on SNF repulsion and out-growth. Norepinephrine and dopamine induced similar growth behavior of SNF. Without treatment,

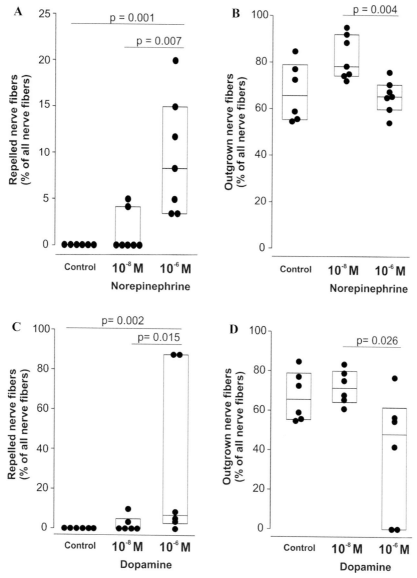

Figure 4. Catecholamine-induced effects on repulsion and growth of sympathetic nerve fibers. (A) Repulsion of sympathetic axons by norepinephrine at different concentrations. (B) Further outgrowth of sympathetic nerve fibers stimulated by norepinephrine at different concentrations. (C) Effects on repulsion of sympathetic axons induced by dopamine. (D) Further outgrowth of sympathetic nerve fibers induced by different doses of dopamine. Each circular area represents one ganglion (observation of growth behavior over 5 hours).

SNFs were not repelled at all (Fig. 4A and C), and approximately 55–85% of nerve fibers grew out further (Fig. 4B and D). Stimulation with norepinephrine at 10^{-8} M repelled few SNF axons (Fig. 4A), and a slightly higher additional outgrowth was recognized compared with controls (Fig. 4B). At a concentration of 10^{-6} M, norepinephrine and dopamine induced a median SNF repulsion of 8 and 5%, respec-

tively (Figs. 4A and C). The effect was variable for norepinephrine and dopamine, with a range of SNF repulsion between 0 and 95% (Fig. 4A and C). Further outgrowth of SNFs was similarly variable with these catecholamines (Fig. 4B and D). In addition, further outgrowth of SNFs was lower at 10^{-6} M norepinephrine compared with 10^{-8} M, indicating a differential effect via β- and α-adrenoceptors, respectively (Fig. 4B). This was similar to dopamine, as dopamine at 10^{-6} M inhibited further outgrowth compared with dopamine at 10^{-8} M (Fig. 4D).

Discussion

This study demonstrates that not only specific axon guidance molecules, such as semaphorins, are involved in the process of SNF repulsion and growth, but hormonal factors, such as catecholamines, and 17β-estradiol or cytokines, such as TNF-α, can also play a role.

The loss of SNFs and their anti-inflammatory neurotransmitters in synovial tissue of RA patients was linked to increased inflammation, and should be considered a proinflammatory signal.[1] This process can lead to an unfavorable situation of chronic autoimmune inflammation and has been interpreted as an evolutionarily conserved mechanism to ensure an adequate inflammatory response, which would be prevented in the presence of SNFs.[26] Although sympathetic neurotransmitters are chemotactic for many immune cells, these activated cells generate a milieu conducive to nerve fiber repulsion. Recently, we immunohistochemically identified semaphorin 3C and 3F in synovial tissue of patients with RA and OA.[3,15] Earlier results revealed that semaphorin 3F alone, though specific to SNF repulsion, is not responsible for the complete disappearance of nerve fibers. Thus, there may be additional factors with repellent activity.

When considering chronic inflammatory diseases like RA, high levels of proinflammatory cytokines in inflamed areas are relevant. Since anti-TNF-α therapy is successfully used, one might assume that this proinflammatory mediator also influences nerve fiber repulsion in the synovium. Our data show that TNF-α is indeed capable of completely repelling SNFs, although this effect was not observed in all experiments. The reasons for this variability are presently not known.

During the last few years it has become evident that not only SNFs but also immune cells are able to produce catecholamines.[19–21] They exert their effects depending on the receptor subtypes and with opposing intracellular signaling pathways.[7] Stimulation of α2-adrenoceptors or activation of D2, D3, and D4 dopamine receptors decreases cyclic AMP and stimulates TNF-α release, whereas binding to β-adrenoceptors or D1 and D5 dopamine receptors increases cAMP and inhibits TNF-α secretion.[27] These locally produced catecholamines might also influence loss of SNFs, but this has not been tested.

In this study, we were able to demonstrate that catecholamines repel SNFs to a certain extent. The effects were highly variable, where, for example, some dopamine experiments resulted in similarly high repulsion rates compared with semaphorin 3F. One can speculate that dopamine and norepinephrine together might potentiate the individual effects, but some preliminary experiments have shown that repulsion of nerve fibers was not enhanced in the presence of both catecholamines (unpublished observations). Similarly, it is not known whether there are any additive or potentiating effects of semaphorins with catecholamines.

The role of estrogens in rheumatoid arthritis were discussed here because both pro- and anti-inflammatory effects have been reported.[28] Estrogens bind to estrogen receptor β (ER-β) and ER-α, which are expressed in synovial fibroblasts and macrophages.[29,30] Estrone sulfate, the most abundant circulating estrogen in nonpregnant women and men, is preferentially converted into estrone and 17β-estradiol in synovial cells.[31] 17β-estradiol seems to exert time- and dose-dependent effects on cell growth and apoptosis.[32] We were interested in the possible effects of 17β-estradiol on SNF repulsion based on the observation that, in the human and rodent uterus, nerve fiber density is related to sex hormone concentration, with a reduction of SNF density at estrous levels.[33] In our present study, a slight repellent and growth-promoting effect of 17β-estradiol was observed *in vitro*. In the future, testing of downstream estrogen that is converted from 17β-estradiol into, for example, 16-hydroxy-estrogens and others might also prove beneficial.

In conclusion, the SNF outgrowth assay enables study of the influence of pathogenetically relevant factors that may be involved in nerve fiber repulsion and growth. We observed that not only specific axon guidance molecules, such as semaphorins, but also other factors, such as catecholamines, 17β-estradiol, and TNF-α, which are elevated in inflamed tissue, influence SNF repulsion and growth. These findings demonstrate a certain redundancy in SNF repulsion that might play an important role in inflamed tissue.

Acknowledgment

This study was supported by grants from the Deutsche Forschungsgemeinschaft (DFG) (Research Unit FOR696).

Conflicts of interest

The authors declare no conflicts of interest.

References

1. Miller, L.E. *et al.* 2000. The loss of sympathetic nerve fibers in the synovial tissue of patients with rheumatoid arthritis is accompanied by increased norepinephrine release from synovial macrophages. *FASEB J.* **14**: 2097–2107.

2. Grum, F. *et al.* 2007. Differential loss of sympathetic nerves and increase of substance P positive nerves in the colon in Crohn's disease: a role for the nerve repellent factor semaphorin 3C. *Brain Behav. Immun.* **20**: e29.

3. Koeck, F.X. *et al.* 2009. Marked loss of sympathetic nerve fibers in chronic Charcot foot of diabetic origin compared to ankle joint osteoarthritis. *J. Orthop. Res.* **27**: 736–741.

4. Mei, Q. *et al.* 2002. Early, selective, and marked loss of sympathetic nerves from the islets of BioBreeder diabetic rats. *Diabetes* **51**: 2997–3002.

5. Lehner, B. *et al.* 2008. Preponderance of sensory versus sympathetic nerve fibers and increased cellularity in the infrapatellar fat pad in anterior knee pain patients after primary arthroplasty. *J. Orthop. Res.* **26**: 342–350.

6. Elenkov, I.J. *et al.* 2000. The sympathetic nervous system—an integrative interface between two supersystems: the brain and the immune system. *Pharmacol. Rev.* **52**: 595–638.

7. Watling, K.J. 1998. *The RBI handbook of receptor classification and signal transduction.* Research Biochemical Inc. Natick, MA.

8. Spengler, R.N. *et al.* 1994. Endogenous norepinephrine regulates tumor necrosis factor-alpha production from macrophages *in vitro. J. Immunol.* **152**: 3024–3031.

9. Spengler, R.N. *et al.* 1990. Stimulation of alpha-adrenergic receptor augments the production of macrophage-derived tumor necrosis factor. *J. Immunol.* **145**: 1430–1434.

10. Fassold, A. *et al.* 2009. Soluble neuropilin-2, a nerve repellent receptor, is increased in rheumatoid arthritis synovium and aggravates sympathetic fiber repulsion and arthritis. *Arthritis Rheum.* **60**: 2892–2901.

11. Yazdani, U. & J.R. Terman. 2006. The semaphorins. *Genome Biol.* **7**: 211.

12. Dickson, B.J. 2002. Molecular mechanisms of axon guidance. *Science* **298**: 1959–1964.

13. Chen, H. *et al.* 1998. Axon guidance mechanisms: semaphorins as simultaneous repellents and anti-repellents. *Nat. Neurosci.* **1**: 436–439.

14. Mark, M.D. *et al.* 1997. Patterning neuronal connections by chemorepulsion: the semaphorins. *Cell Tissue Res.* **290**: 299–306.

15. Miller, L.E. *et al.* 2004. Increased prevalence of semaphorin 3C, a repellent of sympathetic nerve fibers, in the synovial tissue of patients with rheumatoid arthritis. *Arthritis Rheum.* **50**: 1156–1163.

16. Kolodkin, A.L. 1998. Semaphorin-mediated neuronal growth cone guidance. *Prog. Brain Res.* **117**: 115–132.

17. Barker, V. *et al.* 2001. TNFalpha contributes to the death of NGF-dependent neurons during development. *Nat. Neurosci.* **4**: 1194–1198.

18. Hurst, S.M. & S.M. Collins. 1994. Mechanism underlying tumor necrosis factor-alpha suppression of norepinephrine release from rat myenteric plexus. *Am. J. Physiol.* **266**: G1123–G1129.

19. Capellino, S. *et al.* 2010. Catecholamine-producing cells in the synovial tissue during arthritis: modulation of sympathetic neurotransmitters as new therapeutic target. *Ann. Rheum. Dis.* **69**: 1853–1860.

20. Musso, N.R. *et al.* 1996. Catecholamine content and in vitro catecholamine synthesis in peripheral human lymphocytes. *J. Clin. Endocrinol. Metab.* **81**: 3553–3557.

21. Cosentino, M. *et al.* 1999. Endogenous catecholamine synthesis, metabolism, storage and uptake in human neutrophils. *Life Sci.* **64**: 975–981.

22. Clarke, G.L. *et al.* 2010. Beta-adrenoceptor blockers increase cardiac sympathetic innervation by inhibiting autoreceptor suppression of axon growth. *J. Neurosci.* **30**: 12446–12454.

23. Cutolo, M. & R. Wilder. 2000. Different roles for androgens and estrogens in the susceptibility to autoimmune rheumatic diseases. *Rheum. Dis. Clin. North Am.* **26**: 825–839.

24. Zoubina, E.V. & P.G. Smith. 2002. Distributions of estrogen receptors alpha and beta in sympathetic neurons of female rats: enriched expression by uterine innervation. *J. Neurobiol.* **52**: 14–23.

25. Zoubina, E.V. & P.G. Smith. 2003. Expression of estrogen receptors alpha and beta by sympathetic ganglion neurons projecting to the proximal urethra of female rats. *J. Urol.* **169**: 382–385.

26. Straub, R.H. *et al.* 2010. Energy regulation and neuroendocrine-immune control in chronic inflammatory diseases. *J. Intern. Med.* **267**: 543–560.

27. Pivonello, R. *et al.* 2007. Novel insights in dopamine receptor physiology. *Eur. J. Endocrinol.* **156**(Suppl 1): S13–S21.

28. Straub, R.H. 2007. The complex role of estrogens in inflammation. *Endocr. Rev.* **28**: 521–574.

29. Tamir, S. *et al.* 2002. The effect of oxidative stress on ERalpha and ERbeta expression. *J. Steroid Biochem. Mol. Biol.* **81**: 327–332.

30. Cutolo, M. *et al.* 1993. Presence of estrogen-binding sites on macrophage-like synoviocytes and CD8+, CD29+, CD45RO+ T lymphocytes in normal and rheumatoid synovium. *Arthritis Rheum.* **36**: 1087–1097.

31. Schmidt, M. *et al.* 2009. Estrone/17beta-estradiol conversion to, and tumor necrosis factor inhibition by, estrogen metabolites in synovial cells of patients with rheumatoid arthritis and patients with osteoarthritis. *Arthritis Rheum.* **60**: 2913–2922.

32. Cutolo, M. *et al.* 2005. Sex hormone modulation of cell growth and apoptosis of the human monocytic/macrophage cell line. *Arthritis Res. Ther.* **7**: R1124-R1132.

33. Latini, C. *et al.* 2008. Remodeling of uterine innervation. *Cell Tissue Res.* **334**: 1–6.

Ann. N.Y. Acad. Sci. ISSN 0077-8923

ANNALS OF THE NEW YORK ACADEMY OF SCIENCES

Issue: *Neuroimmunomodulation in Health and Disease*

Glucocorticoid–catecholamine interplay within the composite thymopoietic regulatory network

Ivan Pilipović,[1] Katarina Radojević,[1] Milica Perišić,[1] and Gordana Leposavić[2]

[1]Immunology Research Centre "Branislav Janković," Institute of Virology, Vaccines and Sera "Torlak," Belgrade, Serbia.
[2]Department of Physiology, Faculty of Pharmacy, University of Belgrade, Belgrade, Serbia

Address for correspondence: Gordana Leposavić, Ph.D., M.D., Department of Physiology, Faculty of Pharmacy, University of Belgrade, 450 Vojvode Stepe, 11221 Belgrade, Serbia. Gordana.Leposavic@pharmacy.bg.ac.rs

This paper highlights the multiple putative thymic and extrathymic points of intersection and interaction between glucocorticoids (GCs) and catecholamines (CAs)—the end-point mediators of the major routes of communication between the brain and the immune system—in the context of intricate thymic T cell–developmental tuning. More specifically, we discuss in detail findings indicating that adrenal GCs can influence thymopoiesis by adjusting directly and/or indirectly (through modulation of pituitary and local ACTH synthesis) not only thymic GC synthesis, in a cell type–specific manner, but also thymic CA bioavailability (via altering CA outflow from sympathetic nerve endings and local CA synthesis), β and α_1-adrenoceptor (AR) expression, and/or AR-mediated intracellular signal transduction in thymic cells. In addition, this short review points to GC- and CA-sensitive stages along the multistep T cell–developmental journey and the possible effects of altered GC, and consequently CA signaling, on thymopoietic efficiency.

Keywords: glucocorticoids; catecholamines; β-adrenergic receptors; rat thymus; T cell development

Introduction

During the last few decades there has been an exponential increase in both experimental and clinical data illustrating that the nervous and the immune systems are not disparate entities, but systems that crosstalk extensively via multiple pathways.[1–3] The marked influence of psychological stress, as well as depression and other psychiatric disorders, on immune system function has been extensively reported.[4–9] Furthermore, there is evidence that the central nervous system (CNS) can modulate immune cell development and the various stages of an immune response, from the uptake of antigen by antigen-presenting cells or macrophages to proliferation, T cell differentiation, and activity of T or B cells.[10–15] In reverse, immune system changes have been implicated in the pathogenesis of psychiatric disorders such as anxiety or depression.[16–18] Both neural and humoral pathways are involved in relating information between the nervous and the immune systems.[12,19,20] The brain influences immune function by signaling to target cells of the immune system primarily through direct sympathetic innervation of primary and secondary lymphoid organs and through hypothalamic–pituitary–adrenal axis hormonal outflow.[12,19–26] Cytokines released by immune cells signal neuroendocrine, autonomic, limbic, and cortical areas of the CNS to affect neural activity and modify behavior, hormone release, and autonomic function.[21–24] In addition, afferent neural pathways may relay signals from lymphoid and inflamed tissue to the brain.[25,26] Moreover, many signaling molecules (including neurotransmitters, neuropeptides, and hormones) and their specific receptors, which were initially thought to be endogenous to nervous and endocrine systems, have been found in the immune system and vice versa, thereby indicating that a signaling molecule can arise from multiple sources within these three systems to act at multiple targets.[27,28] These findings have laid an important foundation for our understanding that the body's major adaptive systems (nervous, endocrine,

doi: 10.1111/j.1749-6632.2012.06623.x

 Ann. N.Y. Acad. Sci. 1261 (2012) 34–41 © 2012 New York Academy of Sciences.

and immune systems) are in fact interconnected in an extremely complex fashion, allowing the body to adapt to a variety of changes in both internal and external environments. It has been suggested that communication between nervous, endocrine, and immune systems does not consist of linear pathways involving step-by-step cascades of signaling molecules, but most likely consists of intricate network-like communications in which signaling molecules are connected to one another in multiple ways and at multiple levels. Given that the thymus, the primary lymphoid organ responsible for T cell development, is purported to be the key node in this intricate communication,[13,29,30] the complex neuroendocrine–immune signaling network will be dissected at the thymic level. Considering that thymic cells express specific receptors for the end-point hypothalamic–pituitary–adrenal axis and sympathetic nerve signaling molecules (GCs and CAs),[13,31–37] as well as that both GCs and CAs are synthesized in the thymus,[13,38–45] the putative interactions between these signaling molecules in the context of T cell differentiation/maturation tuning will be discussed in detail.

Outline of T cell differentiation/maturation

The thymus is the primary lymphoid organ providing the microenvironment for the complex multistep process of blood-borne thymocyte precursor differentiation/maturation. These cells express neither T cell receptor (TCR) nor CD4 or CD8 molecules and are called *double-negative* (DN) cells. Their maturation involves a productive rearrangement of the TCRβ locus (β-selection) followed by upregulation of CD4 and CD8 gene expression and TCRα locus rearrangement.[46,47] The subsequent CD4+CD8+ double-positive (DP) thymocytes expressing TCRαβ at low level (TCRαβ^{low}) are scrutinized for their ability to recognize self-peptides in the context of self-MHC expressed on thymic nonlymphoid cells. Only potentially useful cells transducing moderate TCRαβ signal strength (positive selection) continue to differentiate into mature CD4+CD8− or CD4−CD8+ single-positive (SP) TCRαβ^{high} cells that emigrate to the periphery. All other cells terminate differentiation either by being eliminated by negative selection (potentially harmful cells transducing a strong TCRαβ signal) or by dying through "neglect" (useless cells).

Glucocorticoids in the thymopoietic regulatory network

It has been shown that apart from thymocytes,[31–33] thymic epithelial and dendritic cells express the functional classical cytosol GC receptor (GR).[34–37] In support of this functionality are data showing high-dose GC administration induces thymic atrophy, affecting not only thymocytes, but also thymic epithelial cells.[35,38,48,49] In the same vein are data indicating that adrenalectomy-induced thymic hypertrophy cannot be reversed by epinephrine (normally produced by the adrenal medulla).[39–41] Our laboratory showed that adrenalectomy led to thymic hypertrophy by increasing the size of the DP TCRαβ$^{−/low}$ thymocyte subset, most likely due to diminished apoptotic elimination of these cells.[42] In support this finding are data that, notwithstanding the fact that DP thymocytes express the lowest GR level when compared with thymocytes at other developmental stages,[32] this thymocyte subset is the most sensitive to GC-induced apoptosis.[43] This apparent discrepancy is related to a study showing that, upon ligand binding, the GR in DP cells can directly translocate to mitochondria, where it initiates an apoptotic cascade; thus there is a correlation between GC treatment–related mitochondrial GR load of DP thymocytes and their sensitivity to GC-induced apoptosis.[43] In the absence of GCs, thymocytes that do not recognize self-MHC or that bear TCRαβ with extremely low avidity to MHC/self-peptide (useless thymocytes) are rescued rather than undergo GC-mediated apoptosis.[44,45,47] Thus, following adrenalectomy, an inefficient apoptotic elimination of useless thymocytes might be expected. The greater number of DP TCRαβ^{low} cells entering selection in these rats, without a proportional rise in the numbers of postselected DP TCRαβ^{high} cells[50] and their SP (CD4+CD8− and CD4−CD8+) TCRαβ^{high} descendants,[50] suggested decelerated transition of DP TCRαβ^{low} thymocytes to downstream differentiation/maturational stages and/or exaggerated negative selection that results in fewer potentially useful cells.[42] The latter option is fully consistent with the mutual antagonism theory, which indicates that the absence of GCs is potentially harmful to thymocytes with low-to-moderate avidity for self-MHC.[47] In the absence of GCs, these cells are forced into activation-mediated apoptosis (negative selection) rather than rescued and positively selected, as would occur in the presence of GCs.[45,51,52] Therefore, it is likely that adrenalectomy leads to the impaired selection of useful thymocytes and thymic generation of functional nonself-reactive T cells.

It is important to note that GC synthesis has been demonstrated in thymic epithelial cells.[53–56] Thymocytes entering the selection processes and postselected DP CD69+ cells have also been shown to synthesize GCs.[57–59] Moreover, it has been suggested that the outcome of GC/TCR signal interplay at the thymocyte level depends on local GC synthesis.[60–62] Therefore, changes in T cell differentiantion/maturation in adrenalectomized rats indirectly suggests that circulating GCs modulate thymic GC synthesis.[42] In support of this notion, downregulation of GC synthesis in thymocytes occurs following adrenalectomy.[57] The lack of GCs by itself does not directly influence thymocyte GC synthesis, but rather the adrenalectomy-induced rise in ACTH concentration does.[57] That is, ACTH in high concentrations diminishes thymic expression of mRNA for CYP11B1, the last enzyme in the corticosterone synthetic route, and consequently GC synthesis is inhibited.[57] In the same vein are data showing that adrenalectomy does not influence CYP11B1 in thymocytes from IL-1β/IL-18 double-knockout mice (unable to respond to adrenalectomy with high ACTH levels).[57] In support of this are findings indicating that thymocytes express the ACTH receptors melanocortin receptor type 2 (MCR2) and MCR5;[59,63] and in these cells, ACTH, in a cAMP-dependent fashion, downregulates expression of mRNA encoding a critical enzyme in steroidogenesis and GC synthesis.[57] Given that ACTH synthesis in thymic cells has been suggested,[64,65] the question arises regarding the relationship between pituitary gland–derived and locally synthesized ACTH. There are data indicating that thymic and serum concentrations of ACTH correlate positively.[66] In light of findings that ACTH influences thymocyte GC synthesis only at high concentrations,[57] one could speculate that the downregulation of thymocyte GC synthesis requires increased local ACTH concentration to be superimposed on the high hormone levels provided by its extrathymic production. Collectively, the aforementioned data indicate that adrenal GC deprivation leads to a rise in thymocyte numbers through an ACTH-mediated reduction of GC synthesis in thymocytes and, consequently, less GC synthesis in thymocytes and, consequently, less

efficient elimination of useless thymocytes through death by neglect. On the other hand, in contrast to thymocytes, in thymic epithelial cells a rise in ACTH concentration enhances GC synthesis.[53] To support the opposing effects of ACTH on distinct types of thymic cells are data showing that GC synthesis is differentially regulated by cAMP in intestinal epithelial and adrenocortical cell lines.[67] In the context of the aforementioned findings suggesting increased elimination of thymocytes with low to moderate avidity for self-MHC in adrenalectomized rats,[42] it may be hypothesized that in the absence of systemic GCs, local production of GCs is not sufficient to efficiently rescue the useful thymocytes. When taken together, the data suggest that while adrenal secretion of GCs into the blood coordinates multiple organ systems, their local synthesis in the thymus—a functionally compartmentalized organ—provides high spatial specificity of GC action.

Glucocorticoid–catecholamine communications in the thymopoietic regulatory network

Multiple findings suggest that changes in systemic GC levels may influence T cell development by altering sympathetic outflow to the thymus. First, it was shown that 15 days after adrenalectomy, the concentration of norepinephrine was decreased in the rat hypothalamus,[68] the brain region in which neurons resides that are involved in the regulation of thymic sympathetic nervous fiber activity;[69] and administration of corticosterone to adrenalectomized rats effectively reversed the decrease in the concentration of this neurotransmitter.[68] Second, there is evidence to indicate that in the rat superior cervical ganglion, where noradrenergic fibers innervating the thymus originate,[70] preganglionic nerve stimulation elicited an increase in CA synthesis and activity.[71] Furthermore, it was demonstrated that GCs can also act directly on the superior cervical ganglia to increase CA synthesis.[72] Third, there are data showing that chronic stress associated with a high GC circulating levels increases the density of CA-containing thymic fibers and, most likely, CA content on a per nerve fiber basis.[48] Finally, CAs, acting through β- and α1-ARs, influence thymic cellularity and T cell differentiation/maturation.[73–76] To further complicate our understanding of GCs' influence on CA-mediated tuning of thymopoiesis are findings indicating that thymocytes, as well as thymic epithelial cells and macrophages, express enzymes involved in CA biosynthesis and degradation, and contain substantial amounts of CAs, including norepinephrine.[77,78] Considering that GCs can modulate CA biosynthesis[79,80] and that all thymocytes express GR,[57–59] one may assume that systemic GCs at least influence CA synthesis in thymocytes by tuning thymic GC synthesis. There are no data on GR expression in CA-synthesizing thymic epithelial cells and macrophages. Yet, there are data that GCs may influence not only CA synthesis, but also their release from CA-synthesizing cells.[81] However, it should also be pointed out that ACTH influences CA biosynthesis by GC-independent mechanisms,[82,83] so one could speculate that systemic GCs may influence thymic CA levels not only by direct action, but also by affecting systemic and thymic ACTH levels. At present, there are no data to confirm GCs' influence on CA bioavailability in the thymus (this research is in progress in our laboratory). However, we have demonstrated that β-AR blockade with propranolol, a nonselective beta blocker, led to more pronounced effects on T cell development in previously adrenalectomized rats than in control animals.[84] This is consistent with data showing that *in vivo* coadministration of hydrocortisone (at a concentration that by itself did not alter the cAMP content of the thymus) with the β-AR agonist isoproterenol induced a rapid, dose-dependent increase in the mouse thymic cAMP content and enhanced the effects of isoproterenol on thymic atrophy.[85]

It has been shown that β-AR blockade has a significant impact on T cell differentiation/maturation in adult rodents by affecting TCRαβ-dependent stages of thymocyte development.[73–75,86] In rats subjected to treatment with propranolol for various times (from 4 to 16 consecutive days), the relative proportion of DP TCRαβ^{low} cells entering selection processes was diminished, while the relative proportions of DP TCRαβ^{high} cells that passed positive selection and their SP (CD4$^+$CD8$^-$ and CD4$^-$CD8$^+$ cells) TCRαβ^{high} descendants were increased.[73,74,86] Such a relationship between the thymocyte subsets strongly suggested enhanced positive and/or reduced negative selection and facilitated maturation of the selected cells following β-AR blockade. This was related to an almost twofold

increase in density of Thy-1 (CD90) on the surface of DP TCRαβ$^{low/high}$ thymocytes.[73,86] There are two lines of evidence to support this assumption. First, it has been shown that exogenous cAMP and norepinephrine can decrease steady-state Thy-1 mRNA levels in S49 mouse thymoma cells and murine thymocytes, and that propranolol prevents this decrease.[87,88] Second, in Thy-1$^{-/-}$ mice thymocyte hyperresponsiveness to TCRαβ triggering, and, consequently, exaggerated negative selection and markedly reduced *de novo* production of mature SP cells from their DP precursors, were reported.[89]

In rats adrenalectomized 4 days prior to the beginning of 4-day propranolol treatment, the increase in frequency of the most mature CD4$^-$CD8$^+$ and CD4$^+$CD8$^-$ TCRαβ^{high} SP thymocytes was greater than the same treatment in nonadrenalectomized rats.[84] This most likely reflected not only changes in thymic CA bioavailability, but also alterations at the β-AR level. Corroborating the latter possibility are data indicating that GCs up-regulate β-AR expression and β-AR–mediated CA action.[79,85,90,91] However, given that the rise in the frequency of CD8$^+$CD4$^-$ SP TCRαβ^{high} cells following propranolol treatment was more pronounced in adrenalectomized than in nonadrenalectomized rats, it was assumed that differential effects of propranolol on T cell differentiation/maturation in adrenalectomized and nonadrenalectomized rats could not solely be ascribed to more efficient β-AR blockade in adrenalectomized rats.[84] Considering that α$_1$-AR–mediated signaling was shown to favor thymocyte differentiation/maturation toward a mature CD4$^-$CD8$^+$ phenotype,[76] and that prolonged exposure of cardiac myocytes to norepinephrine led to diminished α$_1$-AR expression,[92] one may speculate that α$_1$-AR upregulation (due to prolonged decrease in norepinephrine bioavaiability in the thymus) led to preferential differentiation/maturation of CD4$^+$CD8$^+$ DP thymocytes toward a CD4$^-$CD8$^+$ TCRαβ^{high} phenotype. Consequently, it may be hypothesized that changes in the level of systemic GCs may affect CA tuning of thymopoiesis, by decreasing not only the central sympathetic tone and sympathetic norepinephrine outflow in thymus and local thymic CA synthesis (most likely by altering thymic ACTH and/or GC synthesis), but also by altering directly and/or indirectly (via altered CA bioavailability) expression of ARs in the receptor subtype-specific manner.

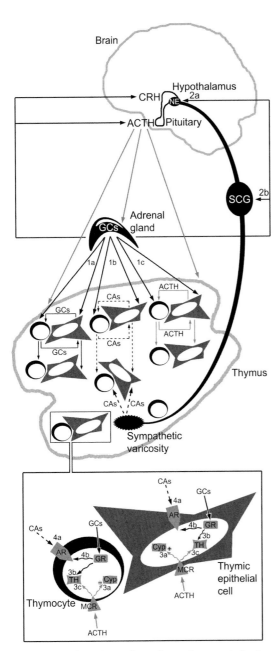

Conclusions and perspectives

There is an increasing body of evidence supporting the concept of communication between the body's major adaptive systems (nervous, endocrine, and immune) as a complex network, wherein a signaling molecule can arise from multiple sources to act at multiple targets, and in which signaling molecules are connected to one another in multiple ways and at multiple levels. In favor of this concept are findings showing an interplay between GCs and CAs in the tuning of thymopoiesis,[84] and in particular those findings pointing to putative points of intersections and crosstalk between GC and CA signaling (Fig. 1). There are data indicating that GCs can affect thymic sympathetic nerve CA outflow, thus influencing activity of the hypothalamic neurons controlling activity of sympathetic nerves innervating the thymus, and the neurons of superior cervical ganglion giving nerve projections to the thymus (Fig. 1).[68,72] In addition, GCs may indirectly (through altering pituitary and possibly thymic ACTH release and thereby local thymic GC synthesis) affect local thymic CA synthesis (Fig. 1). Given that systemic GCs influence GC synthesis in thymocytes and thymic epithelial cells in a reciprocal manner, the question

Figure 1. Schematic outline of putative catecholamine-mediated influence of systemic glucocorticoids on thymopoiesis. (1 and 2) Systemic glucocorticoids (GCs) can influence thymopoiesis by adjusting not only (1a) thymic GC synthesis and consequently GC action on thymocytes and thymic nonlymphoid cells, but also on catecholamine (CA) bioavailability and their modulatory action on thymic lymphoid and nonlymphoid cells. Systemic GCs might influence thymic CA bioavailability by affecting (1b) local CA synthesis and CA outflow from thymic sympathetic nerves by acting at the level of (2a) the hypothalamic norepinephrine-containing neurons that control activity of thymic sympathetic nerves and (2b) the

superior cervical ganglion giving nerve projections to the thymus. In addition, systemic GCs may influence pituitary (directly, or via CRH-dependent mechanisms) and possibly (1c) thymic release of ACTH. (3) ACTH, acting through melanocortin receptors type 2 and type 5, regulates (3a) the expression of CYP11B1 (the last enzyme in GC biosynthesis) in a thymic cell type–specific manner, and consequently (in aaGC-dependent manner) (3b) local CA biosynthesis. In addition, ACTH could exert direct GC-independent influence on (3c) thymic CA biosynthesis. (4) Changes in (4a) CA bioavailability by itself, as well as (4b) by alterations in GC action, could affect the expression of β- and α$_1$-adrenoceptors (ARs) in thymic cells and/or AR-mediated intracellular signal transduction. Note that ACTH, GCs, and CAs synthesized in thymus can possibly act not only in a paracrine but also in an autocrine manner. GCs, glucocorticoids; CAs, catecholamines; NE, norepinephrine; SCG, superior cervical ganglion; CRH, corticotropin-releasing hormone; ACTH, adrenocorticotropic hormone; MCR, melanocortin receptors type 2 and type 5; Cyp, CYP11B1 (the last enzyme in GC biosynthesis); TH, tyrosine hydroxylase (the rate-limiting enzyme in CA synthesis); AR, β- and α$_1$-adrenoceptors; GR, glucocorticoid receptor; black arrow, GC action; gray arrow, ACTH action; dashed arrow, CA action; black-curved arrow, GC-dependent effect; gray-curved arrow, direct ACTH-mediated effect; +, upregulation; −, downregulation.

arises whether alterations in the systemic GC level, and consequently in thymic GC synthesis, differentially influence CA synthesis in distinct thymic cells. In addition, the relationship between sympathetic nerve– and thymic cell–derived CAs remains to be defined. It is worth mentioning that ACTH has been shown to exert a direct GC-independent influence on CA synthesis (Fig. 1).[82,83] There is a strong scientific basis to hypothesize that GCs, acting directly and/or indirectly (via altering thymic CA bioavailability), may also affect expression of distinct AR subtypes and/or AR-mediated intracellular signal transduction, and, consequently, CA tuning of thymopoiesis (Fig. 1). This hypothesis is currently being explored in our laboratory. Finally, it should be pointed out that CAs may also influence efficiency of GC signaling.[93] Understanding GC–CA crosstalk is important not only from a physiological standpoint but also for understanding the mechanisms responsible for thymus involution following exposure to various stressors,[49,94] as well as alterations in thymopoiesis occurring under conditions of altered release of adrenal GCs, due either to disturbances in hypothalamic–pituitary axis activity and chronic administration of GCs or to inhibitors of their synthesis or action under the scope of various therapeutic protocols.

Acknowledgments

This work was supported by Grants 145049 and 175050 from the Ministry of Education and Science of the Republic of Serbia.

Conflicts of interest

The authors declare no conflicts of interest.

References

1. Besedovsky, H.O. & A.D. Rey. 2007. Physiology of psychoneuroimmunology: a personal view. *Brain Behav. Immun.* **21:** 34–44.

2. Eskandari, F. & E.M. Sternberg. 2002. Neural-immune interactions in health and disease. *Ann. N.Y. Acad. Sci.* **966:** 20–27.

3. Straub, R.H. *et al.* 1998. Dialogue between the CNS and the immune system in lymphoid organs. *Immunol. Today* **19:** 409–413.

4. Stein, M., A.H. Miller & R.L. Trestman. 1991. Depression, the immune system, and health and illness. Findings in search of meaning. *Arch. Gen. Psychiatr.* **48:** 171–177.

5. Schleifer, S.J., S.E. Keller & J.A. Bartlett. 1999. Depression and immunity: clinical factors and therapeutic course. *Psychiatr. Res.* **85:** 63–69.

6. Stefanski, V. & H. Engler. 1999. Social stress, dominance and blood cellular immunity. *J. Neuroimmunol.* **94:** 144–152.

7. Schleifer, S.J., S.E. Keller & J.A. Bartlett. 2002. Panic disorder and immunity: few effects on circulating lymphocytes, mitogen response, and NK cell activity. *Brain Behav. Immun.* **16:** 698–705.

8. Segerstrom, S.C. & G.E. Miller. 2004. Psychological stress and the human immune system: a meta-analytic study of 30 years of inquiry. *Psychol. Bull.* **130:** 601–630.

9. Avitsur, R., D.A. Padgett & J.F. Sheridan. 2006. Social interactions, stress, and immunity. *Neurol. Clin.* **24:** 483–491.

10. Maestroni, G.J., A. Conti & E. Pedrinis. 1992. Effect of adrenergic agents on hematopoiesis after syngeneic bone marrow transplantation in mice. *Blood* **80:** 1178–1182.

11. Maestroni, G.J. & A. Conti. 1994. Modulation of hematopoiesis via alpha 1-adrenergic receptors on bone marrow cells. *Exp. Hematol.* **22:** 313–320.

12. Elenkov, I.J. *et al.* 2000. The sympathetic nerve—an integrative interface between two supersystems: the brain and the immune system. *Pharmacol. Rev.* **52:** 595–638.

13. Leposavic, G. *et al.* 2008. Catecholamines as immunomodulators: a role for adrenoceptor-mediated mechanisms in fine tuning of T-cell development. *Auton. Neurosci.* **144:** 1–12.

14. Madden, K.S., V.M. Sanders & D.L. Felten. 1995. Catecholamine influences and sympathetic neural modulation of immune responsiveness. *Annu. Rev. Pharmacol. Toxicol.* **35:** 417–448.

15. Dimitrijevic, M. *et al.* 2009. Chronic propranolol treatment affects expression of adrenoceptors on peritoneal macrophages and their ability to produce hydrogen peroxide and nitric oxide. *J. Neuroimmunol.* **211:** 56–65.

16. Bay-Richter, C. *et al.* 2011. Changes in behaviour and cytokine expression upon a peripheral immune challenge. *Behav. Brain Res.* **222:** 193–199.

17. Nautiyal, K.M. *et al.* 2008. Brain mast cells link the immune system to anxiety-like behavior. *Proc. Natl. Acad. Sci. U.S.A.* **105:** 18053–18057.

18. Dantzer, R. *et al.* 2008. From inflammation to sickness and depression: when the immune system subjugates the brain. *Nat. Rev. Neurosci.* **9:** 46–56.

19. Savastano, S. *et al.* 1994. Hypothalamic–pituitary–adrenal axis and immune system. *Acta Neurol.* **16:** 206–213.

20. Chrousos, G.P. 1995. The hypothalamic–pituitary–adrenal axis and immune-mediated inflammation. *N. Engl. J. Med.* **332:** 1351–1363.

21. Wong, M.L. & E.M. Sternberg. 2000. Immunological assays for understanding neuroimmune interactions. *Arch. Neurol.* **57:** 948–952.

22. Haddad, J.J., N.E. Saade & B. Safieh-Garabedian. 2002. Cytokines and neuro-immune-endocrine interactions: a role for the hypothalamic–pituitary–adrenal revolving axis. *J. Neuroimmunol.* **133:** 1–19.

23. Banks, W.A. 2004. Neuroimmune networks and communication pathways: the importance of location. *Brain Behav. Immun.* **18:** 120–122.

24. Lorton, D. *et al.* 2006. Bidirectional communication between the brain and the immune system: implications for physiological sleep and disorders with disrupted sleep. *Neuroimmunomodulation* **13:** 357–374.

25. Watkins, L.R. *et al.* 1995. Blockade of interleukin-1 induced hyperthermia by subdiaphragmatic vagotomy: evidence for vagal mediation of immune-brain communication. *Neurosci. Lett.* **183:** 27–31.

26. Romeo, H.E. *et al.* 2001. The glossopharyngeal nerve as a novel pathway in immune-to-brain communication: relevance to neuroimmune surveillance of the oral cavity. *J. Neuroimmunol.* **115:** 91–100.

27. Blalock, J.E., D. Harbour-McMenamin & E.M. Smith. 1985. Peptide hormones shared by the neuroendocrine and immunologic systems. *J. Immunol.* **135:** 858s-861s.

28. Blalock, J.E. 1994. The syntax of immune-neuroendocrine communication. *Immunol. Today* **15:** 504–511.

29. Hadden, J.W. 1998. Thymic endocrinology. *Ann. N.Y. Acad. Sci.* **840:** 352–358.

30. Savino, W. & M. Dardenne. 2000. Neuroendocrine control of thymus physiology. *Endocr. Rev.* **21:** 412–443.

31. Wiegers, G.J. *et al.* 2001. CD4(+)CD8(+)TCR(low) thymocytes express low levels of glucocorticoid receptors while being sensitive to glucocorticoid-induced apoptosis. *Eur. J. Immunol.* **31:** 2293–2301.

32. Berki, T. *et al.* 2002. Glucocorticoid (GC) sensitivity and GC receptor expression differ in thymocyte subpopulations. *Int. Immunol.* **14:** 463–469.

33. Boldizsar, F. *et al.* 2010. Emerging pathways of non-genomic glucocorticoid (GC) signalling in T cells. *Immunobiology* **215:** 521–526.

34. Sacedon, R. *et al.* 1999. Glucocorticoid-mediated regulation of thymic dendritic cell function. *Int. Immunol.* **11:** 1217–1224.

35. Talaber, G. *et al.* 2011. Wnt-4 protects thymic epithelial cells against dexamethasone-induced senescence. *Rejuvenation Res.* **14:** 241–248.

36. Riccardi, C., S. Bruscoli & G. Migliorati. 2002. Molecular mechanisms of immunomodulatory activity of glucocorticoids. *Pharmacol. Res.* **45:** 361–368.

37. Pazirandeh, A. *et al.* 2002. Effects of altered glucocorticoid sensitivity in the T cell lineage on thymocyte and T cell homeostasis. *FASEB J.* **16:** 727–729.

38. Dardenne, M., T. Itoh & F. Homo-Delarche. 1986. Presence of glucocorticoid receptors in cultured thymic epithelial cells. *Cell Immunol.* **100:** 112–118.

39. Boinet, B. 1899. Recherches experimentales sur les fonctions des capsules surrenales *CR Seances Soc. Biol. Fil.* **51:** 671–674.

40. Jaffe, H.L. 1924. The influence of the suprarenal gland on the thymus: I. Regeneration of the thymus following double suprarenalectomy in the rat. *J. Exp. Med.* **40:** 325–342.

41. Jaffe, H.L. 1924. The influence of the suprarenal gland on the thymus: III. Stimulation of the growth of the thymus gland following double suprarenalectomy in young rats. *J. Exp. Med.* **40:** 753–759.

42. Stojic-Vukanic, Z. *et al.* 2009. Dysregulation of T-cell development in adrenal glucocorticoid-deprived rats. *Exp. Biol. Med.* **234:** 1067–1074.

43. Talaber, G. *et al.* 2009. Mitochondrial translocation of the glucocorticoid receptor in double-positive thymocytes correlates with their sensitivity to glucocorticoid-induced apoptosis. *Int. Immunol.* **21:** 1269–1276.

44. Chung, H. *et al.* 2002. Rescuing developing thymocytes from death by neglect. *J. Biochem. Mol. Biol.* **35:** 7–18.

45. Ashwell, J.D., L.B. King & M.S. Vacchio. 1996. Cross-talk between the T cell antigen receptor and the glucocorticoid receptor regulates thymocyte development. *Stem Cell.* **14:** 490–500.

46. Zamoyska, R. & M. Lovatt. 2004. Signalling in T-lymphocyte development: integration of signalling pathways is the key. *Curr. Opin. Immunol.* **16:** 191–196.

47. von Boehmer, H. 2004. Selection of the T-cell repertoire: receptor-controlled checkpoints in T-cell development. *Adv. Immunol.* **84:** 201–238.

48. Zivkovic, I. *et al.* 2005. The effects of chronic stress on thymus innervation in the adult rat. *Acta Histochem.* **106:** 449–458.

49. Zivkovic, I.P. *et al.* 2005. Exposure to forced swim stress alters morphofunctional characteristics of the rat thymus. *J. Neuroimmunol.* **160:** 77–86.

50. Shortman, K., D. Vremec & M. Egerton. 1991. The kinetics of T cell antigen receptor expression by subgroups of CD4 + 8+ thymocytes: delineation of CD4 + 8 + 3(2+) thymocytes as post-selection intermediates leading to mature T cells. *J. Exp. Med.* **173:** 323–332.

51. Vacchio, M.S., J.D. Ashwell & L.B. King. 1998. A positive role for thymus-derived steroids in formation of the T-cell repertoire. *Ann. N.Y. Acad. Sci.* **840:** 317–327.

52. Webster, J.I., L. Tonelli & E.M. Sternberg. 2002. Neuroendocrine regulation of immunity. *Annu. Rev. Immunol.* **20:** 125–163.

53. Taves, M.D., C.E. Gomez-Sanchez & K.K. Soma. 2011. Extraadrenal glucocorticoids and mineralocorticoids: evidence for local synthesis, regulation, and function. *Am. J. Physiol. Endocrinol. Metab.* **301:** E11–E24.

54. Lechner, O. *et al.* 2000. Glucocorticoid production in the murine thymus. *Eur. J. Immunol.* **30:** 337–346.

55. Pazirandeh, A. *et al.* 1999. Paracrine glucocorticoid activity produced by mouse thymic epithelial cells. *FASEB J.* **13:** 893–901.

56. Vacchio, M.S., V. Papadopoulos & J.D. Ashwell. 1994. Steroid production in the thymus: implications for thymocyte selection. *J. Exp. Med.* **179:** 1835–1846.

57. Qiao, S., S. Okret & M. Jondal. 2009. Thymocyte-synthesized glucocorticoids play a role in thymocyte homeostasis and are down-regulated by adrenocorticotropic hormone. *Endocrinology* **150:** 4163–4169.

58. Qiao, S. *et al.* 2008. Age-related synthesis of glucocorticoids in thymocytes. *Exp. Cell Res.* **314:** 3027–3035.

59. Gomez-Sanchez, C.E. 2009. Glucocorticoid production and regulation in thymus: of mice and birds. *Endocrinology* **150:** 3977–3979.

60. Wiegers, G.J. *et al.* 2011. Shaping the T-cell repertoire: a matter of life and death. *Immunol. Cell Biol.* **89:** 33–39.

61. Jondal, M., A. Pazirandeh & S. Okret. 2004. Different roles for glucocorticoids in thymocyte homeostasis? *Trends Immunol.* **25:** 595–600.

62. Ashwell, J.D., F.W. Lu & M.S. Vacchio. 2000. Glucocorticoids in T cell development and function*. *Annu. Rev. Immunol.* **18:** 309–345.

63. Johnson, E.W., T.K. Hughes, Jr. & E.M. Smith. 2001. ACTH receptor distribution and modulation among murine mononuclear leukocyte populations. *J. Biol. Regul. Homeost. Agents* **15:** 156–162.

64. Smith, E.M. *et al.* 1990. Nucleotide and amino acid sequence of lymphocyte-derived corticotropin: endotoxin induction of a truncated peptide. *Proc. Natl. Acad. Sci. USA* **87:** 1057–1060.

65. Batanero, E. *et al.* 1992. The neural and neuro-endocrine component of the human thymus: II. Hormone immunoreactivity. *Brain Behav. Immun.* **6:** 249–264.

66. Esquifino, A.I. *et al.* 2001. Age-dependent changes in 24-hour rhythms of thymic and circulating growth hormone and adrenocorticotropin in rats injected with Freund's adjuvant. *Neuroimmunomodulation* **9:** 237–246.

67. Mueller, M. *et al.* 2007. Differential regulation of glucocorticoid synthesis in murine intestinal epithelial versus adrenocortical cell lines. *Endocrinology* **148:** 1445–1453.

68. Rastogi, R.B. & R.L. Singhal. 1978. Evidence for the role of adrenocortical hormones in the regulation of noradrenaline and dopamine metabolism in certain brain areas. *Br. J. Pharmacol.* **62:** 131–136.

69. Trotter, R.N. *et al.* 2007. Transneuronal mapping of the CNS network controlling sympathetic outflow to the rat thymus. *Auton. Neurosci.* **131:** 9–20.

70. Tollefson, L. & K. Bulloch. 1990. Dual-label retrograde transport: CNS innervation of the mouse thymus distinct from other mediastinum viscera. *J. Neurosci. Res.* **25:** 20–28.

71. Biguet, N.F. *et al.* 1989. Preganglionic nerve stimulation increases mRNA levels for tyrosine hydroxylase in the rat superior cervical ganglion. *Neurosci. Lett.* **104:** 189–194.

72. Nagaiah, K., P. MacDonnell & G. Guroff. 1977. Induction of tyrosine hydroxlase synthesis in rat superior cervical ganglia in vitro by nerve growth factor and dexamethasone. *Biochem. Biophys. Res. Commun.* **75:** 832–837.

73. Leposavic, G. *et al.* 2006. Characterization of thymocyte phenotypic alterations induced by long-lasting beta-adrenoceptor blockade in vivo and its effects on thymocyte proliferation and apoptosis. *Mol. Cell Biochem.* **285:** 87–99.

74. Rauski, A. *et al.* 2003. Thymopoiesis following chronic blockade of beta-adrenoceptors. *Immunopharmacol. Immunotoxicol.* **25:** 513–528.

75. Madden, K.S. & D.L. Felten. 2001. Beta-adrenoceptor blockade alters thymocyte differentiation in aged mice. *Cell Mol. Biol.* **47:** 189–196.

76. Leposavic, G. *et al.* 2010. Age-associated plasticity of alpha1-adrenoceptor-mediated tuning of T-cell development. *Exp. Gerontol.* **45:** 918–935.

77. Radojevic, K. *et al.* 2011. Neonatal androgenization affects the efficiency of beta-adrenoceptor-mediated modulation of thymopoiesis. *J. Neuroimmunol.* **239:** 68–79.

78. Pilipovic, I. *et al.* 2008. Sexual dimorphism in the catecholamine-containing thymus microenvironment: a role for gonadal hormones. *J. Neuroimmunol.* **195:** 7–20.

79. Hagerty, T. *et al.* 2001. Interaction of a glucocorticoid-responsive element with regulatory sequences in the promoter region of the mouse tyrosine hydroxylase gene. *J. Neurochem.* **78:** 1379–1388.

80. Fossom, L.H., C.R. Sterling & A.W. Tank. 1992. Regulation of tyrosine hydroxylase gene transcription rate and tyrosine hydroxylase mRNA stability by cyclic AMP and glucocorticoid. *Mol. Pharmacol.* **42:** 898–908.

81. Elhamdani, A. *et al.* 2000. Enhancement of the dense-core vesicle secretory cycle by glucocorticoid differentiation of PC12 cells: characteristics of rapid exocytosis and endocytosis. *J. Neurosci.* **20:** 2495–2503.

82. Serova, L.I. *et al.* 2008. Adrenocorticotropic hormone elevates gene expression for catecholamine biosynthesis in rat superior cervical ganglia and locus coeruleus by an adrenal independent mechanism. *Neuroscience* **153:** 1380–1389.

83. Mueller, R.A., H. Thoenen & J. Axelrod. 1970. Effect of pituitary and ACTH on the maintenance of basal tyrosine hydroxylase activity in the rat adrenal gland. *Endocrinology* **86:** 751–755.

84. Pilipovic, I. *et al.* 2010. Glucocorticoids, master modulators of the thymic catecholaminergic system? *Braz. J. Med. Biol. Res.* **43:** 279–284.

85. Durant, S. 1986. In vivo effects of catecholamines and glucocorticoids on mouse thymic cAMP content and thymolysis. *Cell Immunol.* **102:** 136–143.

86. Pesic, V. *et al.* 2007. Long-term beta-adrenergic receptor blockade increases levels of the most mature thymocyte subsets in aged rats. *Int. Immunopharmacol.* **7:** 674–686.

87. LaJevic, M.D. *et al.* 2010. Thy-1 mRNA destabilization by norepinephrine a 3′ UTR cAMP responsive decay element and involves RNA binding proteins. *Brain Behav. Immun.* **24:** 1078–1088.

88. Wajeman-Chao, S.A. *et al.* 1998. Mechanism of catecholamine-mediated destabilization of messenger RNA encoding Thy-1 protein in T-lineage cells. *J. Immunol.* **161:** 4825–4833.

89. Hueber, A.O. *et al.* 1997. Thymocytes in Thy-1$^{-/-}$ mice show augmented TCR signaling and impaired differentiation. *Curr. Biol.* **7:** 705–708.

90. Abraham, G., G.F. Schusser & F.R. Ungemach. 2003. Dexamethasone-induced increase in lymphocyte beta-adrenergic receptor density and cAMP formation in vivo. *Pharmacology* **67:** 1–5.

91. Cotecchia, S. & A. De Blasi. 1984. Glucocorticoids increase beta-adrenoceptors on human intact lymphocytes in vitro. *Life Sci.* **35:** 2359–2364.

92. Rokosh, D.G. *et al.* 1996. Alpha1-adrenergic receptor subtype mRNAs are differentially regulated by alpha1-adrenergic and other hypertrophic stimuli in cardiac myocytes in culture and in vivo. Repression of alpha1B and alpha1D but induction of alpha1C. *J. Biol. Chem.* **271:** 5839–5843.

93. Corti, A., S. Astancolle & P. Davalli. 1985. Response of hepatic ornithine decarboxylase and polyamine concentration to surgical stress in the rat: evidence for a permissive effect of catecholamines on glucocorticoid action. *Biochem. Biophys. Res. Commun.* **129:** 885–891.

94. Gruver, A.L. & G.D. Sempowski. 2008. Cytokines, leptin, and stress-induced thymic atrophy. *J. Leukoc. Biol.* **84:** 915–923.

Ann. N.Y. Acad. Sci. ISSN 0077-8923

Presentation of neuroendocrine self in the thymus: a necessity for integrated evolution of the immune and neuroendocrine systems

Vincent Geenen

University of Liege, GIGA-Research Center of Immunoendocrinology, Sart Tilman, Belgium

Address for correspondence: Vincent Geenen, M.D., Ph.D., University of Liege, GIGA-Research Center of Immunoendocrinology, CHU-B34, B-4000 Liege-Sart Tilman, Belgium. vgeenen@ulg.ac.be

During evolution, from ancestor thymoids scattered in gill baskets of the lamprey, the first unique thymus appeared in jawed cartilaginous fishes around 450–500 millions years ago, concomitantly or shortly after the emergence of recombinase-dependent adaptive immunity. The major biological function of the thymus is to generate a diverse repertoire of T cell receptors that are self tolerant. The thymus achieves this role by using two complementary and intimately associated mechanisms: apoptotic deletion of T cell clones bearing a TCR with high affinity for self-antigens presented by MHC proteins on thymic epithelial cells (TECs) and dendritic cells (DCs); and generation of self-antigen–specific natural regulatory T (nT_{reg}) cells. Moreover, the escape from thymic central self-tolerance plays a primary role in the development of autoimmune diseases that are a significant burden for the quality of life and health-care cost. Our new knowledge in thymus physiology and physiopathology is currently translated into innovative therapeutic strategies against these devastating chronic diseases.

Keywords. thymus; antigen presentation; central self-tolerance; autoimmunity; AIRE; type 1 diabetes; IGF-2; Graves' disease

The moving place of the thymus in the history of medicine

Claude Galen (129–199 or 217 AD), one of the "fathers" of medicine, and Hippocrates (ca. 460 BC–ca. 370 BC), first reported the observation of the thymus that he so named because of its close resemblance with the leaf of the plant *Thymus vulgaris*. Galen suspected the thymus to be the seat of soul, humor, eagerness, and fortitude. Given Galen's strong influence on Western medicine until the 18th century, this old misconception most probably explains why the words *thymie* and *troubles thymiques* still resonate in French medical language *mood* and *mood disorders* that are observed in neuropsychiatric diseases. Jacopo Berengario da Carpi (1460–1530) then provided the first complete anatomical description of the thymus.

For a long period of time, the thymus was thought to be a vestigial organ that had become useless and redundant during phylogeny and ontogeny after puberty. In the early 1900s, J. August Hammar in Sweden highlighted the important neuroendocrine regulation of the thymus, in particular, the relationship between thymic hyperplasia and acromegaly, Graves' disease, and castration.[1] The thymus was then considered a gland and an intrinsic component of the endocrine system. However, despite the identification of several thymic "hormones," the model of endocrine cell-to-cell signaling failed to characterize the complex molecular dialogue between thymic stromal cells and thymic T cells (thymocytes). In 1959 and 1961, after Jacques F.A.P. Miller demonstrated the crucial role of the thymus in mouse leukemia and T cell development,[2,3] while the endocrine feature of the thymus progressively vanished until the discovery of the intrathymic transcription of neuroendocrine genes.

doi: 10.1111/j.1749-6632.2012.06624.x

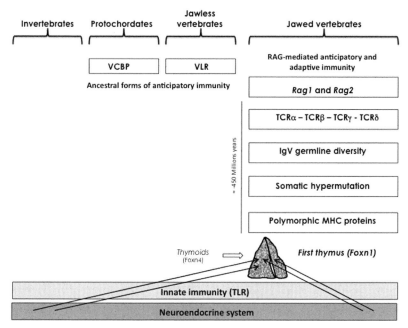

Figure 1. Integrated evolution of the immune and neuroendocrine systems. Essential components of the neuroendocrine system were established long ago and did not display important variation during evolution except for gene duplication and differential RNA splicing. The appearance of RAG-dependent adaptive immunity in jawed vertebrates was associated with a high risk of autotoxicity directed against the neuroendocrine system. Of note, from ancestor lamprey thymoids, the first unique thymus emerged concomitantly in jawed cartilaginous fishes, and the intrathymic presentation of neuroendocrine-related genes (arrows) may be viewed *a posteriori* as a very efficient and economical way to instruct the adaptive T cell system to tolerate neuroendocrine antigens as early as during intrathymic T cell development and differentiation. VCBP, variable-region-containing chitin-binding protein; VLR, variable lymphocyte receptor.

Emergence of the thymus in evolution

In all living species, the neuroendocrine and innate immune systems have evolved in parallel and still coexist today without any problem (Fig. 1). Indeed, Toll-like receptors (TLR), which are the most important mediators of innate immunity, do not have the ability to react against normal and undamaged self. Some 450–500 million years ago, the emergence of transposon-like recombination-activating genes *Rag1* and *Rag2* in jawed fishes (sharks and rays) promoted the development of adaptive immunity.[4–6] The appearance of these elements in the genome of gnathostomes, and the subsequent development of the combinatorial immune system, has been sometimes described as to the "Big Bang" of immunology. Gene recombination in somatic lymphoid cells is responsible for the random generation of the extreme diversity of immune receptors for antigens, B cell (BCR) and T cell receptors (TCR). Because of its inherent autotoxic-

ity, the emergence of this new sophisticated type of immune response exerted an evolutive pressure so strong that, in accordance with Paul Ehrlich's prediction of *horror autotoxicus,* novel structures and mechanisms appeared with the specific role of establishing protection against potential autoimmune attacks to the host (immunological self-tolerance). Of note, the first unique thymus also appeared in the jawed cartilaginous fishes, but was preceded by thymus-like lymphoepithelial structures in the gill baskets of lamprey larvae, as recently demonstrated.[7] These structures named *thymoids* express the gene-encoding forkhead box N4 (*Foxn4*), the orthologue of *Foxn1*, the transcription factor responsible for the differentiation of thymic epithelium in higher vertebrates. Thus, FOXN1 stands at a crucial place in the emergence of thymus epithelium that is an absolute requirement for the control of T cell differentiation and central self-tolerance induction.[8]

Presentation of self in the thymus

Self-antigen presentation by major histocompatibility complex (MHC) proteins on thymic stromal cells (epithelial cells (TECs) and dendritic cells (DCs), mainly) is the central mechanism determining the process of T cell differentiation, which includes three alternative and exclusive fates for developing thymocytes: negative selection of self-reactive T cells generated during the random generation of TCR diversity, catalyzed by recombination-activating enzymes RAG1 and RAG2; selection of self-specific natural regulatory (nT_{reg}) cells; and survival and positive selection of $CD4^+$ and $CD8^+$ effector and self-tolerant T cells. The first two events ensure the establishment of the thymus-dependent central arm of immunological self-tolerance, while the avidity/affinity of the TCR–self-antigen–MHC interaction is the central determinant of T cell negative or positive selection. One important unresolved question, however, is how the same MHC–self-antigen complexes are able to mediate negative selection of self-reactive T cells and yet generate self-specific nT_{reg} cells (extensively discussed in Ref. 9).

Another question has long concerned the nature of self that is presented in the thymus to differentiating T cells, in particular during fetal life. Since its original formulation by Frank M. Burnet, *self* has been a seminal word, first coined in the immunological language as a fecund metaphor with some equivocal correlations to the neurocognitive sciences and even philosophy. The precise identity of self was not elucidated before a series of studies initiated in the late 1980s and 1990s.[10–16] Our personal contribution to this field was to define the biochemical nature of the neuroendocrine self. First, thymic neuroendocrine self-antigens correspond to peptide sequences that are mostly conserved throughout evolution of their related family. Second, a hierarchy and an economic principle characterize their profile of expression in the thymus, as one dominant member per family is synthesized in TECs: that is, oxytocin (OT) for the neurohypophysial family, neurokinin A for tachykinins, neurotensin for neuromedins, corticostatin for somatostatins, and insulin-like growth factor 2 (IGF-2) for the insulin family. This hierarchy is very important because the strength of immunological tolerance to a protein/peptide is proportional to its intrathymic concentration.[17] Third, the autoimmune regulator gene/protein (AIRE) controls the

intrathymic transcription of most of the genes encoding neuroendocrine self-antigens.[18] Following AIRE-regulated gene transcription, thymic neuroendocrine precursors are not processed according to the classic model of neuroendocrine secretion but as antigens for presentation by, or in association with, thymic MHC proteins.[14] Fourth, according to the cryptocrine model of cell-to-cell signaling,[19] thymic T cells express functional neuroendocrine cognate receptors.[20] For example, binding of thymic OT to the OT receptor expressed by thymic pre-T cells phosphorylates focal adhesion kinases,[21] and this might promote the establishment of immunological synapses between TECs and T cells. Finally, for some neuroendocrine-related precursors, their transcription in TECs precedes their eutopic expression in peripheral neuroendocrine glands/cells,[20] and this is also very relevant with regard to the induction of self-tolerance to neuroendocrine principles. Therefore, depending on their behavior as the source of self-antigens or cryptocrine ligands, respectively, the thymic repertoire of neuroendocrine-related precursors transposes at the molecular level the multiple roles of the thymus in T cell differentiation (Fig. 2).[22,23]

The organization of the thymic repertoire of neuroendocrine self-precursors is also significant from an evolutionary point of view. Because neuroendocrine hormones were implicated in the regulation of many physiological functions before the appearance of the anticipatory adaptive immune response, they had to be protected from the risk of autoimmunity inherent to this type of immune lottery. The hypothalamic peptide OT controls different steps of the reproductive process, starting from social affiliation and bonding to parturition and lactation.[24] Consequently, self-tolerance toward OT is important for the preservation of animal and human species. Via its dominant expression in TECs, OT is more tolerated than its homologue vasopressin, which essentially controls water balance and vascular tone. Of note, rare cases of autoimmune diabetes insipidus have been reported,[25–27] whereas autoimmunity toward hypothalamic OT-ergic neurons have never been described. With regard to the insulin family, no autoimmunity has been described against IGF-2, the dominant self-antigen of the insulin family during fetal life, whereas insulin is the primary autoantigen of type 1 diabetes

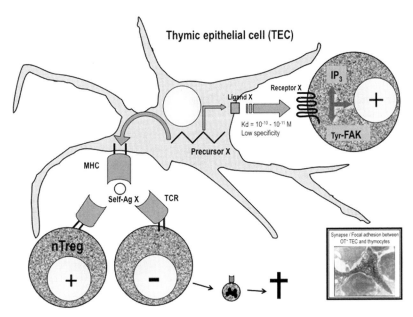

Figure 2. The role of thymic neuroendocrine precursors in T cell differentiation. A precursor X encoded by a neuroendocrine-related gene in a TEC is the source of two distinct types of signaling with thymocytes. First, it delivers a cryptocrine ligand X that is not secreted but targeted at the outer surface of TEC plasma membrane. Through direct membrane-to-membrane contact, this ligand binds with high affinity to a cognate neuroendocrine receptor expressed by thymocytes. For example, OT-mediated cryptocrine signaling activates phosphoinositide turnover with an increase of IP_3 in pre-T cells and phosphorylates focal adhesion-related kinases, which may promote the formation of synapses between TEC and thymocytes. Second, the same precursors may be processed for presentation of neuroendocrine self-epitopes by thymic MHC proteins. Deletion of T cell clones bearing a TCR specific for MHC—neuroendocrine self-antigen complexes, together with generation of self-antigen–specific nT_{reg}, is responsible for the establishment of central self-tolerance toward neuroendocrine gene/protein families. How precisely the same MHC–self-antigen complexes are able to delete self-reactive T cells and select self-specific nT_{reg} cells remains a major unsolved question. FAK, focal adhesion kinase; IP_3, inositol triphosphate.

(T1D).[28] However, through cross-tolerance, thymic neuroendocrine self-antigens promote self-tolerance to all the homologous members of their family as evidenced by the weaker tolerance to insulin in $Igf2^{-/-}$ than in normal mice.[29]

The escape from central self-tolerance as a primary event in autoimmunity, and the concept of *negative self-vaccination*

As already theorized by Burnet, the pathogenesis of autoimmune diseases may first depend on a failure of self-tolerance and the development of "forbidden" self-reactive immune clones.[30] The progressive increase in immune complexity during evolution is associated with a higher incidence of self-tolerance failures, most of them occurring in the human species. There is more and more evidence that a thymus dysfunction in the establishment of central self-tolerance drives the development of the autoimmune response toward many organs. Thymus

transplantation from nonobese diabetic (NOD) mice, an animal model of T1D, was shown to induce diabetes in normal recipients.[31] *Igf2* transcription is deficient in the thymus of diabetes-prone BioBreeding (DPBB) rats, another animal model of T1D, such a defect might contribute to both the absence of tolerance toward β cells and the usual lymphopenia (including $RT6^+$ T_{reg} cells) observed in these animals.[32] Mice with a thymus-restricted insulin defect develop strong proinsulin-specific T cell reactivity,[33] and thymus-specific deletion of insulin induces rapid development of an autoimmune diabetes.[34] Nevertheless, despite the current evidence for a role of thymic insulin in the induction of β cell tolerance (even at a very low level of transcription in medullary TECs), it is important to note that insulin *per se* failed to restore self-tolerance toward β cells in all animal or clinical trials to date.

Loss-of-function *AIRE* single mutations are responsible for a very rare autosomal recessive disease named autoimmune polyendocrinopathy, candidiasis, and ectodermal dystrophy or autoimmune polyglandular syndrome type 1. Depending on their genetic background, *Aire*$^{-/-}$ mice exhibit several signs of peripheral autoimmunity, which are associated with a significant decrease in the intrathymic transcription level of neuroendocrine genes, including those encoding OT, proinsulin 2, and IGF-2.[18,35] Of note, with regard to autoimmune thyroiditis, which is the most frequent autoimmune disease, all major thyroid-related antigens (thyroperoxydase, thyroglobulin, and thyrotropin receptor (TSHR)) are also transcribed in TECs in normal conditions.[15,36] Thymic hyperplasia is commonly observed in Graves' disease,[37] and it was recently shown that homozygotes for an SNP allele predisposing to Graves' disease have significantly lower intrathymic *TSHR* transcripts than carriers of the protective allele.[38] Another credit to a defective central tolerance driving the development of autoimmunity was recently provided with the demonstration of the central role played by a defect in intrathymic α-myosin expression in the pathogenesis of autoimmune myocarditis in mice and humans.[39] Our current in-depth knowledge of thymus physiology and physiopathology has now been translated into the design of innovative tolerogenic and regulatory strategies aimed at restoring central self-tolerance that is absent or defective in autoimmunity.[40–42] The concept of negative self-vaccination has been proposed and is based both on the competition between thymic self-antigens and peripheral target antigens for presentation by MHC proteins, as well as a tolerogenic response—including recruitment of T_{reg} cells—induced by MHC presentation of thymic self-epitopes.[43,44] With this perspective, a research consortium in Wallonia is working on the development of a negative/tolerogenic self-vaccination with the thymic self-peptides related to T1D (Tolediab project).

Thymus involution and immunosenescence

Although thymopoiesis is maintained until late in life,[45–47] thymus involution remains *the* hallmark of immunosenescence that is characterized by a higher susceptibility to infections, as well as a decrease in vaccine and antitumor immune responses. Thymic fat and fibrous involution is associated with a marked decrease in the generation of diverse T cells (in particular, naive CD4$^+$ T cells), an expansion of memory CD8$^+$ T cells, and a diminished influence of central self-tolerance. Involution of the thymus after hypophysectomy was early evidence for the control of the thymus function by a neuroendocrine gland.[48] Since then, numerous studies have unambiguously demonstrated that the hypophysial growth hormone (GH) is able to reverse the age-dependent involution of the thymus.[49–51] Intrathymic proliferation of T cell precursors and thymic output of naive T cells are significantly decreased in adults with GH deficiency, and GH replacement restores these two parameters.[52] Today, restoration of thymus function appears more and more to be an important objective in the elderly, as well as in patients suffering with acquired immunodeficiency syndrome or several hematological diseases.[53,54] It can be anticipated that GH, IGF-1, GH secretagogues (such as ghrelin), GH and ghrelin receptor agonists, as well as other thymus-specific growth factors will be used in the near future for regenerating thymopoiesis and, secondarily, immune functions, including response to vaccines in aged and other immunocompromised patients.

Conclusion

As evidenced in this short overview, a novel era is now beginning for a quantifiable clinical investigation of thymus function in the context of a series of immune-mediated and infectious diseases. Furthermore, pharmacological manipulation of thymus-dependent thymopoietic and tolerogenic functions can now be exploited to provide the scientific community with innovative strategies in the treatment of a large number of immune-mediated disorders.

Acknowledgments

These studies have been supported by the Fund of Scientific Research (F.R.S.-FNRS, Belgium), the Fund for Research in Industry and Agronomy (FRIA, Belgium), the Fund Leon Fredericq for biomedical research at the University Hospital of Liege, the Special Research Fund of the University of Liege, Wallonia (Tolediab, Senegene, ThymUP, and Raparray projects), the Belgian Association of Diabetes, an Independent Research Grant (Pfizer Europe), the European Commission (Eurothymaide FP6 Integrated Project, www.eurothymaide.org), the Juvenile Diabetes Research Foundation (JDRF, New York), and the European Federation for the Study of Diabetes (EFSD, Düsseldorf).

Conflicts of interest

The author declares no conflicts of interest.

References

1. Hammar, J.A. 1921. The new views as to the morphology of the thymus gland and their bearing on the problem of the function of the thymus. *Endocrinology* **5:** 543–573.
2. Miller, J.F.A.P. 1959. Role of the thymus in murine leukaemia. *Nature* **183:** 1069.
3. Miller, J.F.A.P. 1961. Immunological function of the thymus. *Lancet* **2:** 748–749.
4. Agrawal, A., Q.M. Eastman & D.G. Schatz. 1998. Transposition mediated by RAG1 and RAG2 and its implications for the evolution of the immune system. *Nature* **394:** 744–751.
5. Boehm, T. & C. Bleul. 2007. The evolutionary history of lymphoid organs. *Nat. Immunol.* **8:** 131–135.
6. Hirano, M., S. Das, P. Guo & M.D. Cooper. 2011. The evolution of adaptive immunity in vertebrates. *Adv. Immunol.* **109:** 125–157.
7. Bajoghli, B., P. Guo, N. Aghaallaei, *et al.* 2011. A thymus candidate in lampreys. *Nature* **470:** 90–95.
8. Boehm, T. 2011. Design principles of adaptive immune systems. *Nat. Rev. Immunol.* **11:** 307–317.
9. Klein, L., M. Hinterberger, G. Wirnsberger & B. Kyewski. 2009. Antigen presentation in the thymus for positive selection and central tolerance induction. *Nat. Rev. Immunol.* **9:** 833–844.
10. Geenen, V., J.J. Legros, P. Franchimont, *et al.* 1986. The neuroendocrine thymus: coexistence of oxytocin and neurophysin in the human thymus. *Science* **232:** 508–511.
11. Ericsson, A.E., V. Geenen, F. Robert, *et al.* 1990. Expression of preprotachykinin-A and neuropeptide-Y in mRNA of the thymus. *Mol. Endocrinol.* **4:** 1211–1218.
12. Geenen, V., I. Achour, F. Robert, *et al.* 1993. Evidence that insulin-like growth factor 2 (IGF-2) is the dominant thymic member of the insulin superfamily. *Thymus* **21:** 115–127.
13. Jolicœur, C., D. Hanahan & K.M. Smith. 1994. T cell tolerance toward a transgenic beta-cell antigen and transcription of endogenous pancreatic genes in thymus. *Proc. Natl. Acad. Sci. U.S.A.* **91:** 6707–6711.
14. Vanneste, Y., A. Ntodou-Thome, E. Vandersmissen, *et al.* Identification of neurotensin-related peptides in human thymic epithelial cell membranes and relationship with major histocompatibility complex class I molecules. *J. Neuroimmunol.* **76:** 161–166.
15. Sospedra, M., X. Ferrer-Francesch, O. Dominguez, *et al.* 1998. Transcription of a broad range of self-antigens in the thymus suggests a role for central mechanisms in tolerance toward peripheral antigens. *J. Immunol.* **161:** 5918–5929.
16. Derbinski, J., A. Schulte, B. Kyewski & L. Klein. 2001. Promiscuous gene expression in medullary thymic epithelial cells mirrors the peripheral self. *Nat. Immunol.* **2:** 1032–1039.
17. Ashton-Rickardt, P., A. Bandeira, J.R. Delaney, *et al.* 1994. Evidence for a differential avidity model of T cell selection in the thymus. *Cell* **74:** 651–663.
18. Anderson, M.S., E.S. Venanzi, L. Klein, *et al.* 2002. Projection of an immunological self shadow in the thymus by the Aire protein. *Science* **298:** 1395–1401.
19. Funder, J.W. 1990. Paracrine, cryptocrine, acrocrine. *Mol. Cell Endocrinol.* **70:** C21–C24.
20. Hansenne, I., G. Rasier, C. Pequeux, *et al.* 2005. Ontogenesis and functional aspects of oxytocin and vasopressin gene expression in the thymus network. *J. Neuroimmunol.* **158:** 67–75.
21. Martens, H., O. Kecha, C. Charlet-Renard, *et al.* 1998. Neurohypophysial peptides stimulate the phosphorylation of pre-T cell focal adhesions kinases. *Neuroendocrinology* **67:** 282–289.
22. Geenen, V., B. Goxe, H. Martens, *et al.* 1995. Cryptocrine signaling in the thymus network and T cell education to neuroendocrine self-antigens. *J. Mol. Med.* **73:** 449–455.
23. Martens, H., B. Goxe & V. Geenen. 1996. The thymic repertoire of neuroendocrine-related self-antigens: Physiological implications in T cell life and death. *Immunol. Today* **17:** 312–317.
24. Gimpl, G. & F. Fahrenholz. 2001. The oxytocin receptor system: structure, function, and regulation. *Physiol. Rev.* **81:** 629–683.
25. Scherbaum, W.A. & G.F. Bottazzo. 1983. Autoantibodies to vasopressin cells in idiopathic diabetes insipidus: evidence for an autoimmune variant. *Lancet* **1:** 897–901.
26. Imura, H., K. Nakao, A. Shimatsu, *et al.* Lymphocytic infundibuloneurohypophysitis as a cause of central diabetes insipidus. *N. Engl. J. Med.* **239:** 683–689.
27. De Bellis, A., A. Bizzaro & A. Bellastella. 2004. Autoimmune central diabetes insipidus. In *Immunoendocrinology in Health and Disease.* V. Geenen & G.P. Chrousos, Eds.: 439–459. Marcel Dekker. New York.
28. Nakayama, M., N. Abiru, N. Moriyama, *et al.* 2005. Prime role for an insulin epitope in the development of type 1 diabetes in mice. *Nature* **435:** 220–223.
29. Hansenne, I., C. Charlet-Renard, R. Greimers & V. Geenen. 2006. Dendritic cell differentiation and tolerance to insulin-related peptides in *Igf2*-deficient mice. *J. Immunol.* **176:** 4651–4657.
30. Burnet, F.M. 1973. A reassessment of the forbidden clone hypothesis of autoimmune diseases. *Aust. J. Exp. Biol. Med.* **50:** 1–9.
31. Georgiou, H.M. & T.E. Mandel. 1995. Induction of insulitis in athymic (nude) mice. The effect of NOD thymus and pancreas transplantation. *Diabetes* **44:** 49–59.
32. Kecha-Kamoun, O., I. Achour, H. Martens, *et al.* 2001. Thymic expression of insulin-related genes in an animal-model of type 1 diabetes. *Diab. Metab. Res. Rev.* **17:** 146–152.
33. Chentoufi, A. & C. Polychronakos. 2002. Insulin expression levels in the thymus modulate insulin-specific autoreactive T cell tolerance: the mechanism by which the IDDM2 locus may predispose to diabetes. *Diabetes* **41:** 1383–1390.
34. Fan, Y., W.A. Rudert, H. Grupillo, *et al.* 2009. Thymus-specific deletion of insulin induces autoimmune diabetes. *EMBO J.* **28:** 2812–2824.
35. Ramsey, C., O. Winqvist, M. Puhakka, *et al.* Aire-deficient mice develop multiple features of APECED phenotype and show altered immune response. *Hum. Mol. Genet.* **11:** 397–409.
36. Paschke, R. & V. Geenen. 1995. Messenger RNA expression for a TSH receptor variant in the thymus of a two-year old child. *J. Mol. Med.* **73:** 577–580.

37. Murakami, M., Y. Hosoi, T. Negishi, *et al.* 1996. Thymic hyperplasia in patients with Graves' disease. Identification of thyrotropin receptors in human thymus. *J. Clin. Invest.* **98:** 2228–2234.

38. Colobran, R., M. del Pilar Armengol, R. Faner, *et al.* 2011. Association of an SNP with intrathymic transcription of TSHR and Graves' disease: a role for defective thymic tolerance. *Hum. Mol. Genet.* **20:** 3415–3423.

39. Lv, H., E. Havari, S. Pinto, *et al.* 2011. Impaired thymic tolerance to α-myosin directs autoimmunity to the heart in mice and humans. *J. Clin. Invest.* **21:** 1561–1573.

40. Geenen, V., M. Mottet, O. Dardenne, *et al.* 2010. Thymic self-antigens for the design of a negative/tolerogenic self-vaccination gainst type 1 diabetes. *Curr. Opin. Pharmacol.* **10:** 461–472.

41. Chentoufi, A. & V. Geenen. 2011. Thymic self-antigen expression for the design of a negative/tolerogenic self-vaccination against type 1 diabetes. *Clin. Dev. Immunol.* doi:10.11555/2011/349368.

42. Daniel, C., B. Weigmann, R. Bronson & H. von Boehmer. 2011. Prevention of type 1 diabetes in mice by tolerogenic vaccination with a strong insulin mimetope. *J. Exp. Med.* **208:** 1501–1510.

43. Geenen, V. 2006. Thymus-dependent T cell tolerance of neuroendocrine functions. Principles, reflections, and implications for tolerogenic/negative self-vaccination. *Ann. N.Y. Acad. Sci.* **1088:** 284–296.

44. Geenen, V., C. Louis, H. Martens & The Belgian Diabetes Registry. 2004. An insulin-like growth factor 2-derived self-antigen inducing a regulatory cytokine profile after presentation to peripheral blood mononuclear cells from DQ8+ type 1 diabetic adolescents: preliminary design of a thymus-based tolerogenic self-vaccination. *Ann. N.Y. Acad. Sci.* **1037:** 59–64.

45. Douek, D.C., R.D. McFarland, P.H. Keiser, *et al.* 1998. Changes in thymic function with age and during the treatment of HIV infection. *Nature* **396:** 690–695.

46. Poulin, J.F., J.M. Viswanathan, J.M. Harris, *et al.* 1999. Direct evidence for thymic function in adult humans. *J. Exp. Med.* **190:** 479–486.

47. Geenen, V., J.F. Poulin, M.L. Dion, *et al.* 2003. Quantification of T cell receptor rearrangement excision circles to estimate thymic function: an important new tool for endocrine-immune physiology. *J. Endocrinol.* **176:** 305–311.

48. Smith, P. 1930. The effect of hypophysectomy upon the involution of the thymus in rat. *Anat. Rec.* **47:** 119–143.

49. Kelley, K.W., D.A. Weigent & R. Kooijman. 2007. Protein hormones and immunity. *Brain Behav. Immun.* **21:** 384–392.

50. Savino, W. & M. Dardenne. 2000. Neuroendocrine control of thymus physiology. *Endocr. Rev.* **21:** 412–443.

51. Taub, D.D., W.J. Murphy & D.L. Longo. 2010. Rejuvenation of the aging thymus: growth hormone-mediated and ghrelin-mediated signaling pathways. *Curr. Opin. Pharmacol.* **10:** 408–424.

52. Morrhaye, G., H. Kermani, J.J. Legros, *et al.* 2009. Impact of growth hormone (GH) deficiency and GH replacement upon thymus function in adult patients. *PLoS ONE* **4:** e5668.

53. Napolitano, L.A., D. Schmidt, M.B. Gotway, *et al.* 2008. Growth hormone enhances thymic function in HIV-1-infected patients. *J. Clin. Invest.* **118:** 1085–1098.

54. Castermans, E., M. Hannon, J. Durieux, *et al.* 2011. Thymic recovery after allogeneic hematopoietic cell transplantation with non-myeloablative conditioning is limited to patients younger than 60 years of age. *Haematologica* **96:** 298–306.

Ann. N.Y. Acad. Sci. ISSN 0077-8923

ANNALS OF THE NEW YORK ACADEMY OF SCIENCES
Issue: *Neuroimmunomodulation in Health and Disease*

Growth hormone modulates migration of thymocytes and peripheral T cells

Wilson Savino,[1,2] Salete Smaniotto,[2,3] Daniella Arêas Mendes-da-Cruz,[1,2] and Mireille Dardenne[2,4]

[1]Laboratory on Thymus Research, Oswaldo Cruz Institute, Oswaldo Cruz Foundation, Rio de Janeiro, Brazil. [2]Fiocruz-CNRS International Laboratory on Immunology and Immunopathology, Rio de Janeiro, Brazil. [3]Laboratory of Immunohistology, Institute of Biological and Health Sciences, Federal University of Alagoas, Maceió, Brazil. [4]Université Paris Descartes, CNRS UMR-8147, Paris, France

Address for correspondence: Wilson Savino, Laboratory on Thymus Research, Oswaldo Cruz Institute, Oswaldo Cruz Foundation, Ave. Brasil 4365, Manguinhos, Rio de Janeiro, Brazil. savino@fiocruz.br; w_savino@hotmail.com

In the context of immunoneuroendocrine cross talk, growth hormone (GH) exerts pleiotropic effects in the immune system. For example, GH-transgenic mice, as well as animals and humans treated with GH, exhibit enhanced cellularity in the thymus. GH also stimulates the thymic microenvironment, augmenting chemokine and extracellular matrix (ECM) production, with consequent increase in ECM- and chemokine-driven thymocyte migratory responses. Peripheral T cell migration triggered by laminin or fibronectin was enhanced in cells from GH-transgenic versus wild-type control adult mice, as seen for CD4$^+$ and CD8$^+$ T cells from mesenteric lymph nodes. Migration of these T lymphocytes, triggered by the chemokine CXCL12, in conjunction with laminin or fibronectin, was also enhanced compared with control counterparts. Considering that GH can be used as an adjuvant therapy in immunodeficiencies, including AIDS, the concepts defined herein, that GH enhances developing and peripheral T cell migration, provide new clues for future GH-related immune interventions.

Keywords: Growth hormone; thymocytes; cell migration; integrins; chemokines; lymph nodes

Introduction

In the context of crosstalk between the neuroendocrine and immune systems, growth hormone (GH) exerts pleiotropic effects in central as well as peripheral compartments of the immune system.[1,2] For example, in the thymus, a primary lymphoid organ where T cells differentiate, GH upregulates proliferation of distinct cell types, such as thymocytes and thymic epithelial cells. Accordingly, GH-transgenic mice, as well as animals and humans treated with exogenous GH, exhibit an enhanced cellularity in the organ.[3,4] Conversely, there is a severe thymic atrophy in GH receptor-deficient mice.[5] GH also stimulates the secretion of thymic hormones, cytokines, and chemokines by the thymic microenvironment, as well as the production of extracellular matrix (ECM) proteins, including laminin and fibronectin.[1,3] Importantly, these effects are largely mediated by an IGF-1/IGF-1 receptor interaction.[6] Herein, we will summarize recent data on the role of GH upon the migration of thymic and peripheral T lymphocytes.

Migration of developing T lymphocytes is enhanced by GH

From the entry of bone marrow-derived T cell precursors, to the exportation of mature T cells from the thymus to the periphery of the immune system, migration of developing T cells is a key process for normal thymopoiesis.[7]

T cells develop in the thymus from bone marrow–derived precursors that continuously seed the organ via blood vessels. Interestingly, it has been demonstrated in a xenogeneic model that GH stimulates the chemotaxis and the adhesion of human peripheral T cells into the thymus of severe combined immunodeficiency mice, an event partially mediated by integrin-type cell adhesion receptors.[8]

doi: 10.1111/j.1749-6632.2012.06637.x

Within the thymus, the journey of developing thymocytes is controlled by various cell migration-related ligand/receptor pairs, such as those involving ECM proteins and chemokines.[9,10] Accordingly, we postulated that the oriented cell movement of developing thymocytes is a multivectorial process in which each vector corresponds to a given ligand/receptor pair interaction that contributes to the global biological event of cell migration.[11]

Since intratissue lymphocyte migration depends on sequential events of cell adhesion/de-adhesion, we tested whether GH could modulate adhesion of thymocytes to cultured thymic epithelial cells (TEC), the major component of the thymic microenvironmental tridimensional network, which supports the general process of intrathymic T cell differentiation.[12] Treatment of cultured human TEC with GH resulted in increase of thymocyte/TEC adhesion, and the effect was due to the enhancement of the amounts of fibronectin and laminin, together with the expression of their corresponding receptors, VLA-5 and VLA-6, on TEC membranes.[13] Accordingly, when we treated growing TEC with neutralizing antibodies either to ECM ligands or to the corresponding integrin-type receptors, we were able to abrogate the enhancement

of thymocyte adhesion to TEC monolayers. In the same vein, the production of laminin by thymic nurse cells (TNCs, special lymphoepithelial niches in the outer cortex of the thymic lobules) derived from GH-transgenic mice was also increased compared with wild-type controls, and thymocyte release from TNCs was faster in GH-transgenic animals.[3] Moreover, we found that laminin deposition was enhanced in GH-treated normal mice and in GH-transgenic animals, compared with respective controls.[3,14] Nevertheless, it remains to be determined whether the production of various laminin isoforms is upregulated by GH, or if such an effect is restricted to a given isoform. In any case, the responsiveness of thymocytes to laminin-111 is apparent. We showed that in GH-transgenic mice thymocyte adhesion to laminin was higher than in control mice, and an enhancing effect was also observed *ex vivo* when thymocytes were allowed to migrate through laminin-111–coated transwell chambers. The specificity of these effects was demonstrated by showing that they could be blocked with an anti-CD49f monoclonal antibody, which recognizes the alpha-6 chain of the integrin-type laminin receptor VLA-6.[3,14] Interestingly, membrane expression of VLA-6 (which recognizes distinct laminin isoforms) on

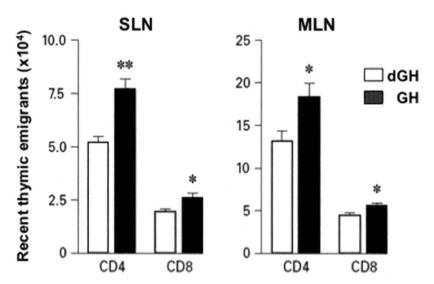

Figure 1. GH treatment enhances *in vivo* homing of CD4$^+$ T cells to lymph nodes. The graphics reveal that normal mice intrathymically treated with GH exhibit more CD4$^+$ and CD8$^+$ recent thymic emigrants in both subcutaneous (SLN) and mesenteric (MLN) lymph nodes, compared with controls injected with heat-induced denatured growth hormone (dGH). In both cases, recent thymic emigrants were defined by cytofluorometry following intrathymic injection of fluoresceinisothiocyanate, harvest of lymph node-derived cells and labeling with anti-CD4 or anti-CD8 fluorochrome-labeled monoclonal antibodies. Statistical significance: $^*P < 0.05$; $^{**}P < 0.01$. Modified from Ref. 27.

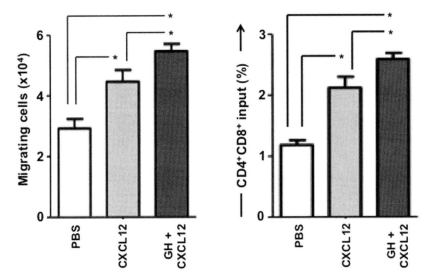

Figure 2. GH enhances CXCL12-induced transendothelial migration of thymocytes. The left panel depicts the absolute numbers of thymocytes that migrated in each experimental condition. The right panel shows the percentage of input for each CD4/CD8-defined thymocyte subset. Transendothelial thymocyte migration was performed in transwell culture inserts. For that, a murine thymic endothelial cell line was cultured onto the inserts of the transwell chambers for 48 hours. Freshly isolated thymocytes (pretreated or not with GH 10^{-8} M for 1 h were then added to thymic endothelial cell monolayer. CXCL12 (100 ng/mL) was applied in the bottom of the transwell chambers. Two million thymocytes were allowed to migrate for 18 h, and migrating cells were harvested in the bottom of the transwell chambers, counted, and phenotyped by flow cytometry. Each bar represents the mean ± standard error of one of three independent experiments. $^*P < 0.05$.

thymocytes from GH-transgenic mice did not change significantly, indicating that GH enhances the activation levels of this integrin rather than increases corresponding gene expression.

Since thymocyte migration is also influenced by chemokines—CXCL12 being one major component—[7,9] we sought to determine whether GH could modulate intrathymic CXCL12-driven migration. In fact, when CXCL12 was applied, thymocyte migration was consistently higher in GH-transgenic mice, compared with age-matched wild-type control animals. This response was blocked by *B. pertussis* toxin, which is known to inhibit various G-protein coupled chemokine receptors. Moreover, we found that when CXCL12 was applied in conjunction with laminin-111, there was a synergic enhancing effect upon migration of GH-transgenic mouse-derived thymocytes. Of note, an increase in CXCL12 production was seen in TEC from GH-transgenic animals, compared with wild-type counterparts, both *in situ* and *in vitro*, as defined at the mRNA and protein levels.[3]

Thymocyte export also seems to be upregulated by GH. We recently evaluated transendothelial migration of GH-treated thymocytes freshly isolated

from normal C57BL/6 mice, in the presence of the chemokine CXCL12 (Fig. 1). When we pretreated thymocytes with GH and led them to migrate toward CXCL12, we observed an increase in the numbers of migrating thymocytes—in particular, those bearing the CD4$^+$CD8$^+$ phenotype—when compared with untreated thymocytes.[4] In addition, *in vivo* experiments showed that GH favors the trafficking of naive CD4$^+$CD8$^-$ recent thymic emigrants to the peripheral lymph nodes, as could be demonstrated in both normal mice treated with GH intrathymically and in GH-transgenic mice,[3,14] as illustrated in Figure 2. Moreover, in acromegalic patients there is an increase in the relative numbers of circulating CD4$^+$ T lymphocytes, compared with age-matched healthy individuals.[15] Of note, an increase in the export of CD4$^+$ T lymphocytes from the thymus was seen in a cohort of HIV-infected patients that received GH as an adjuvant treatment to highly active antiretroviral therapy.[16,17] Accordingly, in these subjects the thymus significantly increased in size, as ascertained *in vivo*.[16]

It has been shown that HIV-positive children with GH-deficiency exhibit diminished thymus volume and reduced circulating CD4$^+$ T cells,

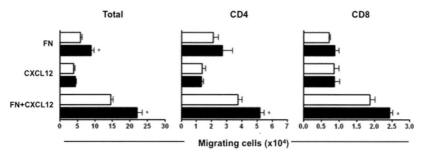

Figure 3. Enhanced migration of mesenteric lymph node derived T cells from GH-transgenic mice. Migration values correspond to the numbers of cells that migrated toward each specific stimulus (fibronectin, CXCL12, or fibronectin combined with CXCL12) subtracted from the values obtained when cells were led to migrate through BSA, applied as an unrelated protein. Experiments were done using two- to three-month-old mice, with at least five animals evaluated per group. The unpaired Student's *t*-test was applied for statistical analyses. Results are expressed as mean ± SEM, and differences between wild-type and transgenic groups were considered statistically significant when *$P < 0.05$.

compared with HIV-positive children without GH deficiency.[18] Taken together, these data show that replenishment of the peripheral pool of CD4+ T lymphocytes can be enhanced by GH-based therapy.

Also, recent data derived from a randomized, placebo-controlled, double-blind study revealed that low-dose GH therapy in highly active antiretroviral therapy (HAART)-treated HIV patients (daily injections of 0.7 mg GH) promoted an increase in thymic emigrants, compared with the placebo group.[19] Moreover, the frequency of recent thymic emigrants, labeled by the detection of T cell receptor excision circles (TREC), and total TREC content significantly increased in the GH group, compared with the placebo group.[19]

GH enhances ECM- and chemokine-driven migratory responses of peripheral T lymphocytes

We recently showed that young and middle-aged GH-Tg mice exhibit higher numbers of B and T lymphocytes in lymph nodes and spleens, compared with wild-type age-matched controls.[20] Similarly, it has been demonstrated that ghrelin, a neuropeptide having a strong GH-releasing activity, induces proliferation of peripheral T lymphocytes.[21] Functionally, we found that migration of T lymphocytes from mesenteric lymph nodes of GH-Tg mice, triggered by the chemokine CXCL12 in conjunction with laminin or fibronectin, was enhanced compared with lymphocytes from wild-type age-matched counterparts[20] (Fig. 3). Since such effects were not correlated with the membrane

densities of the corresponding receptors, these findings indicate that GH enhances the activation state of cell migration–related receptors. It should be noted that the contents of CXCL12, fibronectin, and laminin in peripheral lymphoid organs from GH-transgenic animals were higher than what was found in corresponding control animals, as illustrated in Figure 4. Considering that chemokines attached to the extracellular matrix appear to be better presented to lymphocytes, we can conceive that the efficiency of ligand/receptor pair ligation is higher in a given microenvironment simultaneously enriched in chemokines attached to an enhanced ECM-containing network. These findings indicate that within the lymphoid organs of GH-transgenic mice, T cells should migrate more rapidly, leading to a more rapid recirculation of these lymphocytes.

We also evaluated chemotactic migratory responses driven by the chemokine CCL21, largely known to stimulate migration of peripheral T lymphocytes. We found that CCL21-driven migration of spleen-derived T lymphocytes from GH-transgenic mice, triggered by the chemokine CCL21, was enhanced compared with lymphocytes from control mice. Importantly, such an enhancing effect was even higher when the chemokine was applied in conjunction with the ECM proteins fibronectin or laminin.[20]

Conceptually, these data tell us that the GH-induced increase of migratory responses of lymphocytes in peripheral lymphoid organs is at least partially due to a combined action of selected ECM components and chemokines, thus similar to what we had previously demonstrated for thymocytes.[8,9]

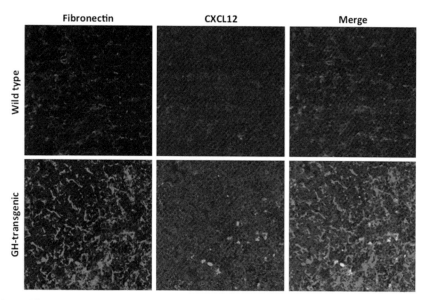

Figure 4. Enhanced deposition of fibronectin and CXCL12 in mesenteric lymph nodes from GH-transgenic mice. The deposition of fibronectin (in red) and CXCL12 (in green) in mesenteric lymph nodes from wild-type and GH-transgenic mice were analyzed by confocal microscopy. Magnification: 400×. Experiments were done using 2- to 3-month-old mice, with three animals being evaluated per group.

In this respect, such data highlight the possibility that the multivectorial concept of cell migration, initially postulated for thymocytes, can be applied for migration of peripheral T lymphocytes within lymphoid tissues, as well as sites of immunological activity. Moreover, the *ex vivo* cell migration data, using fixed concentrations of CXCL12, together with the fact that the presence of such chemokines is actually enhanced within the lymphoid organs suggest that the real *in vivo* effect is likely more important than the one measured *ex vivo*. *In vivo* tracking of these cells will hopefully shed more light on this issue.

Concluding remarks and perspectives

In addition to the well-known role of GH and its secretagogue ghrelin on thymopoiesis in both young and aging individuals,[22] the data summarized here clearly illustrate the role of GH in positively modulating T cell migration, both in the thymus and peripheral lymphoid organs. Yet, from a physiological point of view, there are several issues that deserve to be analyzed. For example, thus far only chemokines and ECM components have been examined as being targets for GH. There are several other interactions involved in the intrathymic trafficking of lymphocytes, including those mediated by semaphorins/neuropilins[23–25]

and Ephs/Ephrins,[26] both having chemorepulsive roles upon thymocyte migration. In addition, nothing is known about whether GH plays a role in a key interaction for thymocyte export, namely, the interaction mediated by sphingosine-1-phosphate and the type 1 sphingosine-1-phosphate receptor.[27]

In any case, and considering that GH can be used as an adjuvant therapy in treating immunodeficiencies, including AIDS,[16,17] the concepts defined herein, that GH enhances migration of both developing and mature peripheral T cells, provide new clues for future GH-related immune interventions.

Acknowledgments

This work was partially funded with grants from Fiocruz, CNPq, Faperj, Fapeal (Brazil), and CNRS (France).

Conflicts of interest

The authors declare no conflicts of interest.

References

1. Savino, W. & M. Dardenne. 2000. Neuroendocrine control of thymus physiology. *Endocr. Rev.* **21:** 412–443.
2. Hattori, N. 2009. Expression, regulation and biological actions of growth hormone (GH) and ghrelin in the immune system. *Growth Hormone IGF Res.* **19:** 187–197.

3. Smaniotto, S., V. de Mello-Coelho, D.M. Villa-Verde, *et al.* 2005. Growth hormone modulates thymocyte development *in vivo* through a combined action of laminin and CXCL12. *Endocrinology* **146:** 3005–3017.

4. Smaniotto, S., A.A. Martins-Neto, M. Dardenne & W. Savino. 2011. Growth hormone is a modulator of lymphocyte migration. *Neuroimmunomodulation* **18:** 309–313.

5. Savino, W., S. Smaniotto, N. Binart, *et al.* 2003. In vivo effects of growth hormone on thymic cells. *Ann. N.Y. Acad. Sci.* **992:** 179–185.

6. Mello-Coelho, V., D.M.S. Villa-Verde, D.A. Farias-de-Oliveira, *et al.* 2002. Functional IGF-1-IGF-1 receptor-mediated circuit in human and murine thymic epithelial cells. *Neuroendocrinology* **75:** 139–150.

7. Ciofani, M. & J.C. Zúñiga-Pflücker. 2007. The thymus as an inductive site for T lymphopoiesis. *Annu. Rev. Cell Dev. Biol.* **23:** 463–493.

8. Taub, D.D., G. Tsarfaty, A.R. Lloyd, *et al.* 1994. Growth hormone promotes human T cell adhesion and migration to both human and murine matrix proteins in vitro and directly promotes xenogeneic engraftment. *J. Clin. Invest.* **94:** 293–300.

9. Savino W., D.A. Mendes-da-Cruz, J.S. Silva, M. Dardenne, *et al.* 2002. Intrathymic T-cell migration: a combinatorial interplay of extracellular matrix and chemokines? *Trends Immunol.* **23:** 305–313.

10. Savino, W., D.A. Mendes-Da-Cruz, S. Smaniotto, *et al.* 2004. Molecular mechanisms governing thymocyte migration: combined role of chemokines and extracellular matrix. *J. Leuk. Biol.* **75:** 1–11.

11. Mendes-da-Cruz, D.A., S. Smaniotto, A.C. Keller, *et al.* 2008. Multivectorial abnormal cell migration in the NOD mouse thymus. *J. Immunol.* **180:** 4639–4647.

12. Petrie, H.T. & J.C. Zúñiga-Pflücker 2007. Zoned out: functional mapping of stromal signaling microenvironments in the thymus. *Annu. Rev. Immunol.* **25:** 649–679.

13. de Mello-Coelho, V., D.M.S. Villa-Verde, M. Dardenne & W. Savino. 1997. Pituitary hormones modulate cell-cell interactions between thymocytes and thymic epithelial cells. *J. Neuroimmunol.* **76:** 39–49.

14. Smaniotto, S., M.M. Ribeiro-Carvalho, M. Dardenne, *et al.* 2004. Growth hormone stimulates the selective trafficking of thymic CD4+CD8- emigrants to peripheral lymphoid organs. *Neuroimmunomodulation* **11:** 299–306.

15. Colao, A., D. Ferone, P. Marzullo, *et al.* 2002. Lymphocyte subset pattern in acromegaly. *J. Endocrinol. Invest.* **25:** 125–128.

16. Napolitano, L.A., J.C. Lo, M.B. Gotway, *et al.* 2002. Increased thymic mass and circulating naive CD4 T cells in HIV-1-infected adults treated with growth hormone. *AIDS* **16:** 1103–1111.

17. Napolitano, L.A., D. Schmidt, M.B. Gotway, *et al.* 2008. Growth hormone enhances thymic function in HIV-1-infected adults. *J. Clin. Invest.* **118:** 1085–1098.

18. Vigano, A., M. Saresella, D. Trabattoni, *et al.* 2004. Growth hormone in lymphocyte thymic and postthymic development: a study in HIV-infected children. *J. Pediatrics* **145:** 542–548.

19. Hansen, B.R., L. Kolte, S.B. Haugaard, *et al.* 2009. Improved thymic index, density and output in HIV-infected patients following low-dose growth hormone therapy: a placebo controlled study. *AIDS* **23:** 2123–2131.

20. Smaniotto, S., D.A. Mendes-da-Cruz, C.E. Carvalho-Pinto, *et al.* 2010. Combined role of extracellular matrix and chemokines on peripheral lymphocyte migration in growth hormone transgenic mice. *Brain Behav. Immun.* **24:** 451–461.

21. Xia, Q., W. Pang, H. Pan, *et al.* 2004. Effects of ghrelin on the proliferation and secretion of splenic T lymphocytes in mice. *Regul. Pept.* **122:** 173–178.

22. Taub, D.D., W.J. Murphy & D.L. Longo. 2010. Rejuvenation of the aging thymus: growth hormone-mediated and ghrelin-mediated signaling pathways. *Curr. Opin. Pharmacol.* **10:** 408–424.

23. Lepelletier, Y., S. Smaniotto, R. Hadj-Slimane, *et al.* 2007. Control of human thymocyte migration by Neuropilin-1/Semaphorin-3A mediated interactions. *Proc. Natl. Acad. Sci. U.S.A.* **104:** 5545–5550.

24. Mendes-da-Cruz, D.A., Y. Lepelletier, A.C. Brignier, *et al.* 2009. Neuropilins, semaphorins, and their role in thymocyte development. *Ann. N.Y. Acad. Sci.* **1153:** 20–28.

25. Garcia, F., Y. Lepelletier, S. Smaniotto, *et al.* 2012. Inhibitory effect of semaphorin-3A, a known axon guidance molecule, in the human thymocyte migration induced by CXCL12. *J. Leukoc. Biol.* **91:** 7–13.

26. Stimamiglio, M.A., E. Jiménez, S.D. Silva-Barbosa, *et al.* 2010. EphB2-mediated interactions are essential for proper migration of T cell progenitors during fetal thymus colonization. *J. Leukoc. Biol.* **88:** 483–494

27. Rosen, H. & E.J. Goetzl. 2005. Sphingosine 1-phosphate and its receptors: an autocrine and paracrine network. *Nat. Rev. Immunol.* **5:** 560–70.

28. Savino, W. & M. Dardenne. 2010. Pleiotropic modulation of thymic functions by growth hormone: from physiology to therapy. *Curr. Opinion Pharmacol.* **10:** 434–442.

Ann. N.Y. Acad. Sci. ISSN 0077-8923

Glucocorticoid regulation of inflammation and its functional correlates: from HPA axis to glucocorticoid receptor dysfunction

Marni N. Silverman and Esther M. Sternberg

Section on Neuroendocrine Immunology and Behavior, National Institute of Mental Health, National Institutes of Health, Bethesda, Maryland

Address for correspondence: Esther M. Sternberg, M.D., Section on Neuroendocrine Immunology and Behavior, National Institute of Mental Health, National Institutes of Health, 10 Center Drive, Rm 2D-39, MSC 1350, Bethesda, MD 20892. sternbee@mail.nih.gov

Enhanced susceptibility to inflammatory and autoimmune disease can be related to impairments in HPA axis activity and associated hypocortisolism, or to glucocorticoid resistance resulting from impairments in local factors affecting glucocorticoid availability and function, including the glucocorticoid receptor (GR). The enhanced inflammation and hypercortisolism that typically characterize stress-related illnesses, such as depression, metabolic syndrome, cardiovascular disease, or osteoporosis, may also be related to increased glucocorticoid resistance. This review focuses on impaired GR function as a molecular mechanism of glucocorticoid resistance. Both genetic and environmental factors can contribute to impaired GR function. The evidence that glucocorticoid resistance can be environmentally induced has important implications for management of stress-related inflammatory illnesses and underscores the importance of prevention and management of chronic stress. The simultaneous assessment of neural, endocrine, and immune biomarkers through various noninvasive methods will also be discussed.

Keywords: stress; cortisol; cytokines; autoimmune; depression; psychoneuroimmunology

Impaired glucocorticoid signaling: from hypothalamic-pituitary-adrenal (HPA) axis to glucocorticoid receptor (GR) dysfunction

Endogenous glucocorticoids play an important role in regulating homeostatic processes under basal and challenge conditions, including metabolism, immune function, and behavior.[1] The essential role of glucocorticoids in protecting the host from the detrimental consequences of an overactive inflammatory immune response has been well established.[2–5] In 1989, we first showed that an impaired HPA axis was an important risk factor for susceptibility to and severity of inflammatory arthritis in autoimmune disease–prone rats.[6] Since that time, impaired HPA responsiveness has been shown in numerous animal models and human inflammatory and autoimmune diseases, including rheumatoid arthritis;[7–10] Crohn's disease, colitis, or inflammatory bowel disease;[11,12] multiple sclerosis and its animal equivalent autoimmune encephalomyelitis;[13] and the allergic conditions, asthma and dermatitis.[14,15] HPA-axis disturbances have also been demonstrated in somatic fatigue and pain disorders such as chronic fatigue syndrome and fibromyalgia,[16–18] and psychiatric disorders such as depression and post-traumatic stress disorder (PTSD),[19–23] which are also associated with an enhanced inflammatory state.[18,24–32] This enhanced inflammatory susceptibility can be related to impairments at any level in the HPA axis, whether at the level of hypothalamic corticotrophin releasing hormone (CRH),[33–35] pituitary adrenocorticotropin (ACTH),[6,36,37] or adrenal glucocorticoid secretion,[6] leading to overall hypocortisolism (Fig. 1); or impairments in local factors affecting glucocorticoid availability and function, including the GR (see below), which can render a state of glucocorticoid resistance by preventing cells and

doi: 10.1111/j.1749-6632.2012.06633.x

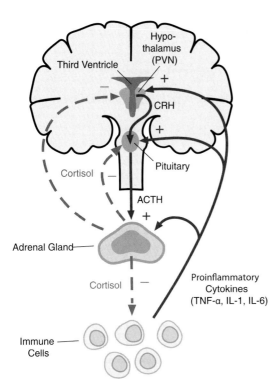

Figure 1. Bidirectional communication between the immune system and the HPA axis. Proinflammatory cytokines, such as TNF-α, IL-1, and IL-6, stimulate glucocorticoid release (cortisol in humans; corticosterone in rodents) by acting at all three levels of the HPA axis (solid blue lines). In turn, glucocorticoids negatively feedback on the immune system to suppress the further synthesis and release of proinflammatory cytokines (dashed red line). In addition, glucocorticoids regulate their own production through negative feedback on the upper levels of the HPA axis, including corticotropin-releasing hormone (CRH) in the paraventricular nucleus (PVN) of the hypothalamus and adrenocorticotropin (ACTH) in the anterior pituitary (dashed red lines). Reprinted with modifications.[98]

tissues in the body from responding adequately to glucocorticoids.

Once glucocorticoids are released into circulation, glucocorticoid availability at the cellular level is influenced by various local factors, including corticosteroid binding globulin (CBG, binding over 90% of circulating glucocorticoids), the multidrug resistance (MDR) P-glycoprotein transporter (an efflux pump that decreases intracellular concentrations of potentially toxic drugs or hormones), and 11β-hydroxysteroid dehydrogenase (11β-HSD, an enzyme with two isoforms: #1 converts inactive glucocorticoids into their active form (e.g., corticosterone in rodents; cortisol in humans); #2 breaks down ac-

tive glucocorticoids into inactive metabolites). Enhanced CBG levels and MDR expression, as well as a decrease in 11β-HSD-1 or increase in 11β-HSD-2, reduce the levels of free/active glucocorticoids in the cell, contributing to a state of glucocorticoid resistance.[10] Glucocorticoid effects are ultimately determined at the level of the GR. An impaired GR, whether as a consequence of reduced expression, binding affinity to its ligand, nuclear translocation, DNA binding, or interaction with other transcription factors (i.e., NF-κB, AP-1), could also lead to a state of glucocorticoid resistance,[38,39] increasing an individual's vulnerability to exaggerated inflammatory responses (Fig. 2). Taken together, even when circulating glucocorticoid concentrations are normal or elevated, impaired counterregulatory control of immune responses can still occur at the cellular and molecular level.

Because glucocorticoid resistance occurs through many different molecular mechanisms, conditions that appear phenotypically similar may have different molecular bases in different individuals or populations. When combined with the many molecular pathways that could potentially disrupt HPA axis hormone production, secretion, availability, and/or receptor function, the number of potential molecular mechanisms that might result in impaired HPA function may seem daunting. This review will focus on impaired GR function as a molecular mechanism of glucocorticoid resistance, which may play a role in the final common pathway of many inflammatory-related conditions.

Genetic and environmental factors can impair GR function

Both genetic and environmental factors can contribute to impaired GR function. Numerous polymorphisms and mutations of the GR have been linked with glucocorticoid resistance and associated with various disease states.[40–42] Alternative splicing of the human GR primary transcript produces multiple isoforms.[43] Increased expression of the GRβ isoform—an inactive form of the GR that competes with the active GRα form—can cause relative glucocorticoid resistance (Fig. 2).[44] Increased expression of GRβ is stimulated by proinflammatory cytokines,[45] and has been found in immune cells of patients with various inflammatory conditions.[10,45] In addition, alternative translation initiation of GR mRNA and various posttranslational

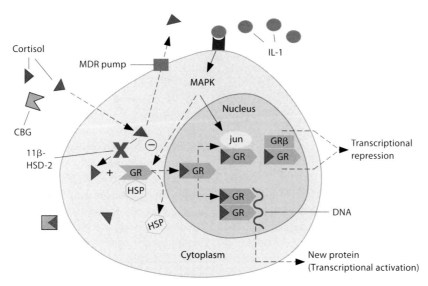

Figure 2. Glucocorticoid resistance can be attributed to changes in local/cellular factors regulating glucocorticoid bioavailability and action. (1) Increased corticosteroid binding globulin (CBG); (2) increased expression of the multidrug resistance transporter (MDR pump); (3) decreased expression of 11β-hydroxy steroid dehydrogenase (HSD)-1 or increased expression of 11β-HSD-2; (4) reduced glucocorticoid binding to the glucocortoid receptor (GRα); (5) reduced GR translocation from the cytoplasm to the nucleus, which could be affected by its phosphorylation state (via mitogen-activated protein kinase [MAPK] pathways); (6) increased GR interaction with inflammatory-related transcription factors, such as NF-κB or AP-1 (jun/fos); and (7) increased GRα interaction with GRβ. HSP, heat shock protein. Reprinted with modifications.[98]

protein modifications can lead to the expression of GR isoforms with different transcriptional activity.[43] Interestingly, GR polymorphisms that confer reduced glucocorticoid sensitivity have been found in both the human GRα and GRβ gene. The A3669G variant of GRβ increases its mRNA stability and dominant negative function,[46] and the ER22/23EK polymorphism in GRα produces higher expression of the GR translational variant GR-A, leading to decreased expression of the more transcriptionally active GR-B protein (see Fig. 1 in Ref. 41 and Fig. 3 in Ref. 43).[47] Epigenetic changes in the GR gene have also been shown as a result of early life behavioral programming in both rats and humans.[48,49]

Other environmental factors that can induce glucocorticoid resistance include chronic inflammation, exposure to infectious agents, chronic exposure to exogenous glucocorticoids, and chronic stress. Cytokines and their downstream signaling molecules can influence the expression and activity of various factors that regulate local glucocorticoid bioavailability and GR function, which are altered in various chronic inflammatory disease states, usually favoring glucocorticoid resistance.[10,50,51] We have

recently shown that exposure to bacterial toxins, including *Bacillus anthracis* lethal toxin and *Clostridia sordelli* lethal toxin, can induce a state of glucocorticoid resistance, with impaired GR function, reduced GR transactivation, and enhanced mortality in mice.[52–54] Chronic exposure to high-dose glucocorticoid therapy also induces glucocorticoid resistance, as in steroid resistant asthma.[15] Chronic psychosocial stress has also been shown to induce glucocorticoid resistance in immune cells in humans and mice.[55–58] Moreover, both psychogenic and immune stressors can induce similar neuroendocrine (HPA-activating) and neurotransmitter changes in the brain, therefore, sensitizing the brain to subsequent stressors (of either type) and, hence, inducing a state of increased stress vulnerability as seen in various psychiatric and somatic disorders.[59] By synergizing, neuroendocrine-immune disturbances present in chronic exposure to either psychogenic or immune stressors can further contribute to the development of glucocorticoid resistance via impaired GR function,[60,61] and therefore produce a vicious feed-forward cycle perpetuating reduced inhibitory feedback on the HPA axis and inflammatory responses.

This suggests a mechanism for the long ob-served phenomenon of severe or chronic stress triggering the onset or exacerbations of inflamma-tory/autoimmune diseases, psychiatric and somatic disorders,[18,50,62] and even metabolic disorders such as metabolic syndrome.[63,64] It would appear coun-terintuitive to predict that stress-related illnesses such as depression, metabolic syndrome, cardio-vascular disease, or osteoporosis have inflammation as a common final pathway, as they are commonly characterized by a hyperactive HPA axis,[65] and the presence of chronically elevated levels of glucocorti-coids should suppress inflammation. However, the induction of glucocorticoid resistance and enhanced inflammation in these states provides a potential mechanism for this effect[21] and may explain their high comorbidity.

Impaired GR function as a mechanism for exaggerated cytokine-induced behavioral and metabolic alterations

Depression is a case in point. Unrestrained in-flammation and HPA axis activity, nonresponsive-ness to suppression by the synthetic glucocorticoid dexamethasone, as well as metabolic alterations, are commonly observed in certain subsets of de-pressed patients and may be due to glucocorticoid resistance via impaired GR function.[19,21,61,66] This phenomenon may be of particular relevance to de-pression in the medically ill, which is more consis-tently characterized by an enhanced inflammatory state.[24,26,27] Innate immune cytokines can influence virtually every pathophysiologic domain relevant to depression, including neuroendocrine function, neurotransmitter metabolism, regional brain activ-ity, and ultimately, behavior.[28,67] Cytokines have been shown to induce a constellation of symp-toms referred to as "sickness behavior," in which an animal's motivational state is shifted toward a behavioral depression (lethargy, reduced locomo-tor activity and food intake, increased sleep) in the effort to conserve energy for fever production and immune activation.[68] Many features of sick-ness behavior overlap with symptoms of depression, particularly neurovegetative symptoms (i.e., psy-chomotor slowing, fatigue, anorexia, weight loss, hypersomnia). Sickness behavior is usually acute, considered adaptive for the recovery from infection, and usually subsides along with the resolution of in-flammation.[68] However, sustained levels of low-level

inflammation (as seen in many chronic inflamma-tory disorders) can contribute to prolonged expres-sion of these behaviors, and in some conditions, cul-minate in the maladaptive development of depres-sion.[24,26–28] Indeed, greater expression of sickness behaviors/neurovegetative symptoms is a promi-nent feature of immunotherapy (e.g., IFN-α)–induced depression (in patients without pretreat-ment depression), compared with depression in medically healthy individuals.[69] The importance of intact glucocorticoid responses in keeping in-flammatory responses and their behavioral seque-lae in check is supported by the observation that adrenalectomized animals or those administered a GR antagonist exhibit exaggerated peripheral and central inflammatory responses, as well as enhanced sickness and depressive-like behaviors to an inflam-matory challenge.[70–74]

Similar to the metabolic effects of glucocor-ticoids (promoting muscle proteolysis, adipose lipolysis, and hepatic gluconeogenesis),[65] proin-flammatory cytokines promote a catabolic state to provide energy sources for increased glucose availability (required for thermogenesis/fever and sustained immune activation).[75,76] Therefore, sus-tained metabolic effects of an enhanced inflam-matory state may contribute to an "inflammatory" metabolic syndrome.[63,64] Moreover, metabolic syn-drome has been associated with depressive symp-tomatology characterized primarily by neurovege-tative features.[77] Although enhanced inflammation appears to represent a link between depressive-like behavior and metabolic alterations, the molecular mechanisms by which an impaired GR may be a vulnerability factor for inflammation-induced be-havioral and metabolic dysfunction requires further elucidation.

The GR is a ligand-dependent transcription fac-tor that belongs to the nuclear hormone receptor super family. The two main molecular pathways by which GR controls transcription are by: (1) dimer-izing and directly binding to positive or negative glucocorticoid response elements in the promoter regions of genes leading to transcriptional increases (i.e., anti-inflammatory and metabolic molecules) or decreases (i.e., negative feedback on the HPA axis), and (2) GR monomers indirectly influencing the transcription of genes through protein–protein interactions with other transcription factors (i.e., inhibiting the transcriptional activity of NF-κB and

AP-1 on proinflammatory molecules; Fig. 2).[38,39] Whereas the metabolic actions of glucocorticoids are believed to be mediated by GR-DNA binding mechanisms, the extent to which GR-DNA binding mechanisms contribute to anti-inflammatory effects of GR remains an issue of debate.[78,79] The predominant role of GR protein–protein interactions with other transcription factors (e.g., NF-κB, AP-1) in mediating GR's anti-inflammatory effects is mainly supported by *in vitro* studies.[39,80,81] Under *in vivo* conditions, the dependence of GR's anti-inflammatory effects on GR-DNA binding-dependent mechanisms has been shown to differ according to the type of immune challenge.[81–84]

To dissect the relative importance of GR-DNA binding-dependent mechanisms versus protein–protein interactions in GR regulation of the sickness syndrome (HPA, inflammatory, behavioral, and metabolic responses) to a systemic inflammatory challenge, we administered low-dose lipopolysaccharide (LPS) to GR^dim mice. GR^dim mice contain a point mutation that renders them deficient in GR dimerization, and hence, GR-DNA binding-dependent mechanisms of transcriptional regulation are impaired, but GR protein–protein interactions with other transcription factors remain intact.[80] We have shown that GR^dim mice administered low-dose LPS exhibit exaggerated peripheral and central inflammatory responses in a cytokine-, time-, and tissue-specific manner. In addition, these mice exhibit sustained LPS-induced plasma corticosterone levels and sickness behaviors, as well as enhanced metabolic alterations and impaired thermoregulation (Silverman, unpublished data). Taken together, these data show that GR dimerization, and hence GR-DNA binding-dependent mechanisms, are essential for the proper regulation and recovery of HPA axis, inflammatory, behavioral, and metabolic responses to a systemic immune challenge such as LPS.

Our data also support the concept that GR protein–protein interactions are not sufficient for glucocorticoids to exert their full anti-inflammatory effects, and suggest that endogenous glucocorticoid responses or exogenous glucocorticoid therapy limited to GR protein–protein interactions could predispose individuals to prolonged behavioral and metabolic sequelae of an enhanced inflammatory state. These include a continued state of sickness/depressive behaviors such as lethargy/fatigue,

reduced locomotor activity, and anorexia, and metabolic alterations favoring a sustained catabolic state (especially increased proteolysis in skeletal muscle required for increased gluconeogenesis in the liver), which can lead to muscle wasting, fat gain (cachectic obesity), insulin resistance (selective inhibition of fuel storage in liver, adipose tissue, and muscle), osteopenia (reduced calcium absorption in bone), and anemia (reduced iron in blood) to provide sustained fuel to the activated immune system.[63] Therefore, many features traditionally attributed to elevated glucocorticoids could actually result from impaired GR function, and hence glucocorticoid resistance and ensuing enhanced inflammation.[21]

This suggests a general principle for a protective role of intact glucocorticoid negative feedback and a fully functioning GR in a range of illnesses in which inflammation plays a central role. And, hence, at the molecular level, the functional status of the GR may play a role in the final common pathway in determining relative vulnerability versus resilience to many inflammatory-related conditions. Further elucidation of the molecular mechanisms involved in GR regulation of inflammation will help inform the pathophysiology of and potential risk factors for clinical conditions associated with HPA axis dysfunction and prolonged inflammation (e.g., autoimmune disease, depression, chronic fatigue, fibromyalgia, PTSD, cardiovascular disease, type 2 diabetes, metabolic syndrome, osteoporosis)[18,21,63,64,67] and may also explain their high comorbidity. Moreover, these insights can better inform the development of glucocorticoid-related therapeutic agents for inflammatory disorders and their downstream correlates.[79,85,86]

Neural-immune biomarker assessment and therapeutic implications

Given the many levels of the HPA axis that can be potentially interrupted, and the combination of environmental and genetic factors that can induce glucocorticoid resistance, it is not surprising that stress-induced inflammatory diseases vary greatly in their presentation in different individuals and populations, and may include features that are inflammatory, behavioral, and/or metabolic related. The evidence that glucocorticoid resistance can be environmentally induced has important implications for management of stress-related inflammatory illnesses and underscores the importance of prevention and management of chronic stress. If the initial goals of research in the fields encompassed by the terms "psychoneuroimmunology," "neuroendocrine immunology," or "neuroimmunomodulation" were to prove and elucidate the connections between the immune and central nervous systems, the goals for the next generation of research in these fields should be to bring this knowledge into the clinic to help prevent and treat stress-related diseases. In this regard, a thorough understanding of the role of glucocorticoid resistance in illness and of environmental factors that modify or prevent it will be essential in taking neuroimmunomodulation to the translational level to prevent and treat human illness.

To do so, new tools will be required to fully characterize the functioning of the central nervous and immune systems at functional, physiological, hormonal, and molecular levels. Ideally, such tools should be noninvasive or minimally invasive and should provide profiles of multiple biomarkers to obtain signature patterns of analytes indicative of the status of these systems. We are currently developing a method to measure neural and immune biomarkers in sweat patch eluates extracted from cutaneous patches worn on the abdomen. Our previous proof-of-principle studies showed that sweat patch analytes (i.e., proinflammatory cytokines and neuropeptides) correlated with plasma levels in healthy controls[87] and with both plasma levels and symptom severity in a population of women with a history of major depressive disorder in remission.[88] Salivary cortisol has long been used to noninvasively assess the circadian rhythm of cortisol.[89] More recently, attention has been paid to the salivary cortisol awakening response as a sensitive measure to detect HPA axis dysregulation.[90] Cortisol measurements in hair provide another promising new method on the horizon to measure the cumulative burden of chronic stress over a longer time period (months).[91] Although we have not discussed the role of the autonomic nervous system (ANS) in neuroimmune interactions[92–95] or HPA–ANS interactions[21] in this review, heart rate variability (HRV) is another minimally invasive method that measures the relative activity of the parasympathetic and sympathetic nervous systems. Decreased HRV, indicative of reduced parasympathetic-vagal tone, is an independent risk factor for morbidity and mortal-

ity.[96] In addition, salivary alpha-amylase measurements have been used as a noninvasive biomarker for the sympathetic nervous system.[97]

These methods coupled with other more invasive methods requiring blood draws will provide a snapshot of an individual's health status and will allow more comprehensive approaches to predicting those at risk and monitoring the effectiveness of therapeutic interventions and coping strategies in stress-related disorders. Ultimately, prediction and management of stress-related illnesses will require a comprehensive work-up, screening for both genetic predisposition factors such as polymorphisms of the GR and related proteins, and functional status of the neuroendocrine stress response, the ANS, and immune/inflammatory responses. Simultaneous assessment of neural, endocrine, and immune biomarkers may help inform the design of more effective therapeutic interventions, whether pharmacological or behavioral, to optimize the reversal of chronic stress' averse effects on the body, normalize stress-induced neuroendocrine-immune disturbances, and restore proper GR function.

Acknowledgment

This work was supported by the NIH, NIMH Division of Intramural Research.

Conflicts of interest

The authors declare no conflicts of interest.

References

1. Sapolsky, R.M., L.M. Romero & A.U. Munck. 2000. How do glucocorticoids influence stress responses? Integrating permissive, suppressive, stimulatory, and preparative actions. *Endocr. Rev.* **21:** 55–89.
2. Besedovsky, H.O. & A. del Rey. 1996. Immune-neuroendocrine interactions: facts and hypotheses. *Endocr. Rev.* **17:** 64–102.
3. McEwen, B.S. *et al.* 1997. The role of adrenocorticoids as modulators of immune function in health and disease: neural, endocrine and immune interactions. *Brain Res. Brain Res. Rev.* **23:** 79–133.
4. Webster, J.I. & E.M. Sternberg. 2004. Role of the hypothalamic-pituitary-adrenal axis, glucocorticoids and glucocorticoid receptors in toxic sequelae of exposure to bacterial and viral products. *J. Endocrinol.* **181:** 207–221.
5. Glezer, I. & S. Rivest. 2004. Glucocorticoids: protectors of the brain during innate immune responses. *Neuroscientist* **10:** 538–552.
6. Sternberg, E.M. *et al.* 1989a. Inflammatory mediator-induced hypothalamic-pituitary-adrenal axis activation is

defective in streptococcal cell wall arthritis- susceptible Lewis rats. *Proc. Natl. Acad. Sci. U.S.A.* **86:** 2374–2378.

7. Crofford, L.J. 2002. The hypothalamic-pituitary-adrenal axis in the pathogenesis of rheumatic diseases. *Endocrinol. Metab. Clin. North Am.* **31:** 1–13.

8. Chikanza, I.C., W. Kuis & C.J. Heijnen. 2000. The influence of the hormonal system on pediatric rheumatic diseases. *Rheum. Dis. Clin. North Am.* **26:** 911–925.

9. Straub, R.H. *et al.* 2005. How psychological stress via hormones and nerve fibers may exacerbate rheumatoid arthritis. *Arthritis Rheum.* **52:** 16–26.

10. Silverman, M.N. & E.M. Sternberg. 2008. Neuroendocrine-immune interactions in rheumatoid arthritis: mechanisms of glucocorticoid resistance. *Neuroimmunomodulation* **15:** 19–28.

11. Stasi, C. & E. Orlandelli. 2008. Role of the brain-gut axis in the pathophysiology of Crohn's disease. *Dig. Dis.* **26:** 156–166.

12. Mawdsley, J.E. & D.S. Rampton. 2005. Psychological stress in IBD: new insights into pathogenic and therapeutic implications. *Gut* **54:** 1481–1491.

13. Gold, S.M. *et al.* 2005. The role of stress-response systems for the pathogenesis and progression of MS. *Trends Immunol.* **26:** 644–652.

14. Buske-Kirschbaum, A. 2009. Cortisol responses to stress in allergic children: interaction with the immune response. *Neuroimmunomodulation* **16:** 325–332.

15. Adcock, I.M. *et al.* 2008. Steroid resistance in asthma: mechanisms and treatment options. *Curr. Allergy Asthma Rep.* **8:** 171–178.

16. Parker, A.J., S. Wessely & A.J. Cleare. 2001. The neuroendocrinology of chronic fatigue syndrome and fibromyalgia. *Psychol. Med.* **31:** 1331–1345.

17. Neeck, G. & L.J. Crofford. 2000. Neuroendocrine perturbations in fibromyalgia and chronic fatigue syndrome. *Rheum. Dis. Clin. North Am.* **26:** 989–1002.

18. Silverman, M.N. *et al.* 2010. Neuroendocrine and immune contributors to fatigue. *Physical Med. Rehab.* **2:** 338–346.

19. Holsboer, F. 2000. The corticosteroid receptor hypothesis of depression. *Neuropsychopharmacology* **23:** 477–501.

20. Heim, C., U. Ehlert & D.H. Hellhammer. 2000. The potential role of hypocortisolism in the pathophysiology of stress-related bodily disorders. *Psychoneuroendocrinology* **25:** 1–35.

21. Raison, C.L. & A.H. Miller. 2003. When not enough is too much: the role of insufficient glucocorticoid signaling in the pathophysiology of stress-related disorders. *Am. J. Psychiatry* **160:** 1554–1565.

22. Yehuda, R. 2009. Status of glucocorticoid alterations in post-traumatic stress disorder. *Ann. N.Y. Acad. Sci.* **1179:** 56–69.

23. Anacker, C. *et al.* 2011. The glucocorticoid receptor: pivot of depression and of antidepressant treatment? *Psychoneuroendocrinology* **36:** 415–425.

24. Pollak, Y. & R. Yirmiya. 2002. Cytokine-induced changes in mood and behaviour: implications for 'depression due to a general medical condition', immunotherapy and antidepressive treatment. *Int. J. Neuropsychopharmacol.* **5:** 389–399.

25. Maes, M. 2008. Inflammatory and oxidative and nitrostative stress pathways underpinning chronic fatigue, somatization and psychosomatic symptoms. *Curr. Opin. Psychiatry* **22:** 75–83.

26. Maes, M. *et al.* 2009. The inflammatory and neurodegenerative (I&ND) hypothesis of depression: leads for future research and new drug developments in depression. *Metab. Brain Dis.* **24:** 27–53.

27. Raison, C.L., L. Capuron & A.H. Miller. 2006. Cytokines sing the blues: inflammation and the pathogenesis of depression. *Trends Immunol.* **27:** 24–31.

28. Dantzer, R. *et al.* 2008. From inflammation to sickness and depression: when the immune system subjugates the brain. *Nat. Rev. Neurosci.* **9:** 46–57.

29. Loftis, J.M., M. Huckans & M.J. Morasco. 2010. Neuroimmne mechanisms of cytokine-induced depression: current theories and novel treatment strategies. *Neurobiol. Dis.* **37:** 519–533.

30. Pace, T.W. & C.M. Heim. 2011. A short review on the psychoneuroimmunology of posttraumatic stress disorder: from risk factors to medical comorbidities. *Brain Behav. Immun.* **25:** 6–13.

31. Gill, J.M. *et al.* 2009. PTSD is associated with an excess of inflammatory immune activities. *Perspect. Psychiatr. Care* **45:** 262–277.

32. Gur, A. & P. Oktayoglu. 2008. Status of immune mediators in fibromyalgia. *Curr. Pain Headache Rep.* **12:** 175–181.

33. Sternberg, E.M. *et al.* 1989b. A central nervous system defect in biosynthesis of corticotropin-releasing hormone is associated with susceptibility to streptococcal cell wall-induced arthritis in Lewis rats. *Proc. Natl. Acad. Sci. U.S.A.* **86:** 4771–4775.

34. Zelazowski, P. *et al.* 1993. Release of hypothalamic corticotropin-releasing hormone and arginine-vasopressin by interleukin 1 beta and alpha MSH: studies in rats with different susceptibility to inflammatory disease. *Brain Res.* **631:** 22–26.

35. Calogero, A.E. *et al.* 1992. Neurotransmitter-induced hypothalamic-pituitary-adrenal axis responsiveness is defective in inflammatory disease-susceptible Lewis rats: in vivo and in vitro studies suggesting globally defective hypothalamic secretion of corticotropin-releasing hormone. *Neuroendocrinology* **55:** 600–608.

36. Zelazowski, P. *et al.* 1992. In vitro regulation of pituitary ACTH secretion in inflammatory disease susceptible Lewis (LEW/N) and inflammatory disease resistant Fischer (F344/N) rats. *Neuroendocrinology* **56:** 474–482.

37. Bernardini, R. *et al.* 1996. Adenylate-cyclase-dependent pituitary adrenocorticotropin secretion is defective in the inflammatory-disease-susceptible Lewis rat. *Neuroendocrinology* **63:** 468–474.

38. Barnes, P.J. 2006. Corticosteroid effects on cell signalling. *Eur. Respir J.* **27:** 413–426.

39. De Bosscher, K., W. Vanden Berghe & G. Haegeman. 2006. Cross-talk between nuclear receptors and nuclear factor kappaB. *Oncogene* **25:** 6868–6886.

40. van Rossum, E.F. & E.L. van den Akker. 2010. Glucocorticoid resistance. *Endocr. Dev.* **20:** 127–136.

41. Manenschijn, L. *et al.* 2009. Clinical features associated with glucocorticoid receptor polymorphisms. An overview. *Ann. N.Y. Acad. Sci.* **1179:** 179–198.

42. DeRijk, R. & E.R. de Kloet. 2005. Corticosteroid receptor genetic polymorphisms and stress responsivity. *Endocrine* **28:** 263–270.

43. Zhou, J. & J.A. Cidlowski. 2005. The human glucocorticoid receptor: one gene, multiple proteins and diverse responses. *Steroids* **70:** 407–417.

44. Lewis-Tuffin, L.J. & J.A. Cidlowski. 2006. The physiology of human glucocorticoid receptor beta (hGRbeta) and glucocorticoid resistance. *Ann. N.Y. Acad. Sci.* **1069:** 1–9.

45. Webster, J.C. *et al.* 2001. Proinflammatory cytokines regulate human glucocorticoid receptor gene expression and lead to the accumulation of the dominant negative beta isoform: a mechanism for the generation of glucocorticoid resistance. *Proc. Natl. Acad. Sci. U.S.A.* **98:** 6865–6870.

46. Derijk, R.H. *et al.* 2001. A human glucocorticoid receptor gene variant that increases the stability of the glucocorticoid receptor beta-isoform mRNA is associated with rheumatoid arthritis. *J. Rheumatol.* **28:** 2383–2388.

47. Russcher, H. *et al.* 2005. Increased expression of the glucocorticoid receptor-A translational isoform as a result of the ER22/23EK polymorphism. *Mol. Endocrinol.* **19:** 1687–1696.

48. Weaver, I.C. *et al.* 2004. Epigenetic programming by maternal behavior. *Nat. Neurosci.* **7:** 847–854.

49. McGowan, P.O. *et al.* 2009. Epigenetic regulation of the glucocorticoid receptor in human brain associates with childhood abuse. *Nat. Neurosci.* **12:** 342–38.

50. Marques, A.H., M.N. Silverman & E.M. Sternberg. 2009. Glucocorticoid dysregulations and their clinical correlates. From receptors to therapeutics. *Ann. N.Y. Acad. Sci.* **1179:** 1–18.

51. Pace, T.W., F. Hu & A.H. Miller. 2007. Cytokine-effects on glucocorticoid receptor function: relevance to glucocorticoid resistance and the pathophysiology and treatment of major depression. *Brain Behav. Immun.* **21:** 9–19.

52. Tait, A.S. *et al.* 2007. The large clostridial toxins from Clostridium sordellii and C. difficile repress glucocorticoid receptor activity. *Infect. Immun.* **75:** 3935–3940.

53. Webster, J.I. & E.M. Sternberg. 2005. Anthrax lethal toxin represses glucocorticoid receptor (GR) transactivation by inhibiting GR-DNA binding in vivo. *Mol. Cell Endocrinol.* **241:** 21–31.

54. Webster, J.I. *et al.* 2003. Anthrax lethal factor represses glucocorticoid and progesterone receptor activity. *Proc. Natl. Acad. Sci. U.S.A.* **100:** 5706–5711.

55. Rohleder, N. 2012. Acute and chronic stress induced changes in sensitivity of peripheral inflammatory pathways to the signals of multiple stress systems—2011 Curt Richter Award Winner. *Psychoneuroendocrinology* **37:** 307–316.

56. Miller, G.E. *et al.* 2008. A functional genomic fingerprint of chronic stress in humans: blunted glucocorticoid and increased NF-kappaB signaling. *Biol. Psychiatry* **64:** 266–272.

57. Miller, G.E., S. Cohen & A.K. Ritchey. 2002. Chronic psychological stress and the regulation of proinflammatory cytokines: a glucocorticoid-resistance model. *Health Psychol.* **21:** 531–541.

58. Avitsur, R. *et al.* 2009. Social interactions, stress, and immunity. *Immunol. Allergy Clin. North Am.* **29:** 285–293.

59. Hayley, S., Z. Merali & H. Anisman. 2003. Stress and cytokine-elicited neuroendocrine and neurotransmitter sensitization: implications for depressive illness. *Stress* **6:** 19–32.

60. Barden, N. 2004. Implication of the hypothalamic-pituitary-adrenal axis in the physiopathology of depression. *J. Psychiatry Neurosci.* **29:** 185–193.

61. Zunszain, P.A. *et al.* 2011. Glucocorticoids, cytokines and brain abnormalities in depression. *Prog. Neuropsychopharmacol Biol. Psychiatry* **35:** 722–729.

62. Sternberg, E.M. 2006. Neural regulation of innate immunity: a coordinated nonspecific host response to pathogens. *Nature Rev. Immunol.* **6:** 318–328.

63. Straub, R.H. *et al.* 2010. Energy regulation and neuroendocrine-immune control in chronic inflammatory diseases. *J. Intern. Med.* **267:** 543–560.

64. Black, P.H. 2003. The inflammatory response is an integral part of the stress response: implications for atherosclerosis, insulin resistance, type II diabetes and metabolic syndrome X. *Brain Behav. Immun.* **17:** 350–364.

65. Chrousos, G.P. & T. Kino. 2007. Glucocorticoid action networks and complex psychiatric and/or somatic disorders. *Stress* **10:** 213–219.

66. McIntyre, R.S. *et al.* 2007. Should depressive syndromes be reclassified as "metabolic syndrome type II"? *Ann. Clin. Psychiatry* **19:** 257–264.

67. Miller, A.H., V. Maletic & C.L. Raison. 2009. Inflammation and its discontents: the role of cytokines in the pathophysiology of major depression. *Biol. Psychiatry* **65:** 732–741.

68. Hart, B.L. 1988. Biological basis of the behavior of sick animals. *Neurosci. Biobehav. Rev.* **12:** 123–137.

69. Capuron, L. *et al.* 2009. Does cytokine-induced depression differ from idiopathic major depression in medically healthy individuals? *J. Affect. Disord.* **119:** 181–185.

70. Goujon, E. *et al.* 1997. Regulation of cytokine gene expression in the central nervous system by glucocorticoids: mechanisms and functional consequences. *Psychoneuroendocrinology* **22:** S75–S80.

71. Nadeau, S. & S. Rivest. 2003. Glucocorticoids play a fundamental role in protecting the brain during innate immune response. *J. Neurosci.* **23:** 5536–5544.

72. Johnson, R.W., M.J. Propes & Y. Shavit. 1996. Corticosterone modulates behavioral and metabolic effects of lipopolysaccharide. *Am. J. Physiol.* **270:** R192–R198.

73. Pezeshki, G., T. Pohl & B. Schöbitz. 1996. Corticosterone controls interleukin-1 beta expression and sickness behavior in the rat. *J. Neuroendocrinol.* **8:** 129–135.

74. Wang, D. *et al.* 2011. Chronic blockade of glucocorticoid receptors by RU486 enhances lipopolysaccharide-induced depressive-like behaviour and cytokine production in rats. *Brain Behav. Immun.* **25:** 706–714.

75. Beisel, W.R. 1975. Metabolic response to infection. *Annu. Rev. Med.* **26:** 9–20.

76. Chang, H.R. & B. Bistrian. 1998. The role of cytokines in the catabolic consequences of infection and injury. *JPEN J. Parenter. Enteral. Nutr.* **22:** 156–166.

77. Capuron, L. *et al.* 2008. Depressive symptoms and metabolic syndrome: is inflammation the underlying link? *Biol. Psychiatry* **64:** 896–900.

78. Clark, A.R. 2007. Anti-inflammatory functions of glucocorticoid-induced genes. *Mol. Cell. Endocrinol.* **275:** 79–97.

79. Schacke, H., W.D. Docke & K. Asadullah. 2002. Mechanisms involved in the side effects of glucocorticoids. *Pharmacol. Ther.* **96:** 23–43.

80. Reichardt, H.M. *et al.* 1998. DNA binding of the glucocorticoid receptor is not essential for survival. *Cell* **93:** 531–541.

81. Reichardt, H.M. *et al.* 2001. Repression of inflammatory responses in the absence of DNA binding by the glucocorticoid receptor. *EMBO J.* **20:** 7168–7173.

82. Grose, R. *et al.* 2002. A role for endogenous glucocorticoids in wound repair. *EMBO Rep.* **3:** 575–582.

83. Tuckermann, J.P. *et al.* 2007. Macrophages and neutrophils are the targets for immune suppression by glucocorticoids in contact allergy. *J. Clin. Invest.* **117:** 1381–1390.

84. Kleiman, A. *et al.* 2012. Glucocorticoid receptor dimerization is required for survival in septic shock via suppression of interleukin-1 in macrophages. *FASEB J.* **26:** 722–729.

85. De Bosscher, K. *et al.* 2008. Selective transrepression versus transactivation mechanisms by glucocorticoid receptor modulators in stress and immune systems. *Eur. J. Pharmacol.* **583:** 290–302.

86. Kleiman, A., & J.P. Tuckermann. 2007. Glucocorticoid receptor action in beneficial and side effects of steroid therapy: lessons from conditional knockout mice. *Mol. Cell. Endocrinol.* **275:** 98–108.

87. Marques-Deak, A. *et al.* 2006. Measurement of cytokines in sweat patches and plasma in healthy women: validation in a controlled study. *J. Immunol. Methods* **315:** 99–109.

88. Cizza, G. *et al.* 2008. Elevated neuroimmune biomarkers in sweat patches and plasma of premenopausal women with major depressive disorder in remission: the POWER study. *Biol. Psychiatry* **64:** 907–911.

89. Kirschbaum, C. & D.H. Hellhammer. 1994. Salivary cortisol in psychoneuroendocrine research: recent developments and applications. *Psychoneuroendocrinology* **19:** 313–333.

90. Clow, A., F. Hucklebridge & L. Thorn. 2010. The cortisol awakening response in context. *Int. Rev. Neurobiol.* **93:** 153–175.

91. Dettenborn, L. *et al.* 2012. Introducing a novel method to assess cumulative steroid concentrations: increased hair cortisol concentrations over 6 months in medicated patients with depression. *Stress* **15:** 348–353.

92. Elenkov, I.J. *et al.* 2000. The sympathetic nerve—an integrative interface between two supersystems: the brain and the immune system. *Pharmacol. Rev.* **52:** 595–638.

93. Nance, D.M. & V.M. Sanders. 2007. Autonomic innervation and regulation of the immune system. *Brain Behav. Immun.* **21:** 736–745.

94. Tracey, K.J. 2007. Physiology and immunology of the cholinergic anti-inflammatory pathway. *J. Clin. Invest.* **117:** 289–296.

95. Thayer, J.F. 2009. Vagal tone and the inflammatory reflex. *Cleve. Clin. J. Med.* **76:** S23–S26.

96. Thayer, J.F. & E. Sternberg. 2006. Beyond heart rate variability: vagal regulation of allostatic systems. *Ann. N.Y. Acad. Sci.* **1088:** 361–372.

97. Nater, U.M. & N. Rohleder. 2009. Salivary alpha-amylase as a non-invasive biomarker for the sympathetic nervous system: current state of research. *Psychoneuroendocrinology* **34:** 486–496.

98. Silverman, M.N., B.D. Pearce & A.H. Miller. 2003. Cytokines and HPA axis regulation. In *Cytokines and Mental Health*. Z. Kronfol, Ed.: 85–122. Kluwer Adademic Publishers. Norwell, MA.

Ann. N.Y. Acad. Sci. ISSN 0077-8923

Hsp72, inflammation, and aging: causes, consequences, and perspectives

Eduardo Ortega,[1] María Elena Bote,[1] Hugo Oscar Besedovsky,[2] and Adriana del Rey[2]

[1]Department of Physiology, University of Extremadura, Badajoz, Spain. [2]Department of Immunophysiology, Institute of Physiology and Pathophysiology, Marburg, Germany

Address for correspondence: Dr. Eduardo Ortega, Department of Physiology (Group of Immunophysiology), Faculty of Sciences, University of Extremadura, Avda de Elvas s/n 06071-Badajoz, Spain. orincon@unex.es

Although aging is an inexorable component of life, its progress depends on how cumulative disruptions of homeostasis are compensated. Cumulative oxidative and inflammatory processes must be controlled to maintain successful aging. Heat shock proteins, such as those of the Hsp70 family, can be considered a danger signal, and their effects can either support longevity by neutralizing danger or can become detrimental when their production is not balanced. Here, we discuss evidence indicating that these highly conserved proteins can favor longevity when such balance is achieved. We emphasize mechanisms affected by Hsp72 that can interfere with effects of excessive oxidative stress and subtle inflammation and, acting either directly or by affecting neuro-immune-endocrine interactions, can mediate metabolic, neuroprotective, and behavioral adjustments during the aging process.

Keywords: Hsp70; longevity; stress; innate response; neuro-immune-endocrine interactions

Introduction

Heat shock proteins (Hsp), also termed *stress proteins*, are highly conserved molecules found in all cellular organisms (prokaryotes and eukaryotes). The primary role of Hsp is to chaperone, transport, and fold proteins when cells are exposed to different types of stress. Under physiological conditions, Hsp are expressed at low levels. However, a wide variety of pathological and physiological stressful stimuli can induce a marked increase in the synthesis of intracellular Hsp, a process known as the cellular stress response.[1] The Hsp70 family constitutes the most conserved and best-studied class of Hsp. It includes constitutively expressed Hsp70 (Hsc70; 73 kDa) and stress-inducible Hsp70 (Hsp72; 72 kDa). The synthesis of intracellular Hsp72 increases markedly following stressful stimuli (stress response), and the protein acts as an intracellular molecular chaperone protecting the cells from further stress factors.[2–4] In addition, it has been shown that Hsp72 can be released *in vitro* by different immune cell types.[5–8] It is also present in the circulation in healthy individuals,[9,10] and at elevated levels in a number of pathological states, including infection and inflammation.[5,11–13] Sensory/motor stress also results in the release of Hsp72 into the bloodstream in both humans and experimental animals,[3,14] and NA is involved in the systemic release of this protein.[7,15]

Physical stress–induced circulating concentration of extracellular Hsp72 (eHsp72) promotes chemotaxis, phagocytosis, and the fungicidal capacity of neutrophils. eHsp72 also stimulates the release of proinflammatory cytokines by monocytes and macrophages, and their capacity to phagocytize, process, and present antigens. These effects involve the participation of Toll-like receptors (mainly TLR2 and TLR4) and adrenoreceptors; NF-κB and cAMP, among others, act as intracellular signals that mediate eHsp72 signaling.[16–21] These findings suggested that eHsp72 plays an important role in the initiation of adaptive immune responses and in innate immune mechanisms during stressful situations. Thus, these molecules are considered to be a danger signal.[2,11,22]

doi: 10.1111/j.1749-6632.2012.06619.x

Hsp72 and aging

The immune system interacts with the nervous and endocrine systems in a network-like fashion, and extrinsic signals are also involved in immunoregulation. Neurotransmitters, neurohormones, hormones, and cytokines act as messengers within this neuro-endocrine-immune network and mediate an exchange of information about the actual state of the different components of the systems involved. This immune-neuro-endocrine network is constantly in operation and performs adjustments of homeostasis necessary to support host defenses and to restore health.[23] There is also evidence that the neuro-endocrine-immune communication is disrupted during the aging process, leading to increased morbidity and mortality.[24,25] Although immunodepression has been considered a component of aging, there is now a large body of evidence suggesting that aging does not equally affect all cells of the immune system. Immunosenescence (i.e., age-related changes of the immune system) was first related to a pronounced decrease in T helper functions, which affects humoral immunity and causes an impaired B cell response.[26,27] However, aging differentially affects the cells that participate in innate and/or inflammatory responses. Particularly for macrophages, aging does not always induce a decline in their functions.[28,29] For example, it has been reported that peripheral blood mononuclear cells from aged people produce and release higher amounts of proinflammatory cytokines than those of young people.[30] Peritoneal macrophages from old animals also have a greater phagocytic activity and aerobic microbicidal capacity (as evaluated by superoxide anion production), compared with those from young counterparts, and neuro/sensorial "stress mediators," such as corticosterone and noradrenaline, are involved in the magnitude of this activation.[29,31,32] Also, the expression of Toll-like receptors, such as TLR2 and TLR4, by peritoneal leukocytes increases with aging.[33] The different effect of aging on cells involved in adaptive immunity and macrophages and other cells responsible for natural immunity probably reflects a physiological adaptation that could counterbalance the decreased lymphoid activity in aged subjects,[28] but likely at the cost of promoting a low-grade chronic inflammation that is termed *inflammaging*.[34–36] Oxidative stress is a key feature in aging,[37] and there is a close link between oxidation and inflammation.[38] On this basis, the hypothesis of *oxi-inflammation* was proposed by De la Fuente and Miquel (a concept that integrates the oxidation and inflammaging processes and assumes that aging is a form of chronic oxidative and inflammatory stress that leads to cell damage and an age-related decline in the functions of several regulatory systems).[24] Free radicals and inflammatory mediators exert a microbicide activity and contribute to an optimal defensive function, but their production can be deregulated during the aging process. Thus, successful aging and longevity also depend on efficient anti-inflammatory and antioxidative mechanisms.[24,36,39]

Most human studies indicate that circulating Hsp72 levels decrease with advancing age in a normal population.[40–45] Because inflammation induces the release of Hsp72, it has been proposed that low serum Hsp72 levels reflect an anti-inflammatory status associated with successful biological aging and longevity.[36,41,44,45] This view is supported by studies in centenarians and their offspring.[43,45] In this context, low circulating Hsp72 levels have been recently proposed as a "marker of longevity."[45] However, the fact that basal levels of Hsp72 are reduced in centenarians does not necessarily mean that this protein is detrimental, because there are multiple lines of evidence indicating that Hsp72 exerts scavenging effects, protects against the oxidative process in different cells and tissues, and exerts neuroprotection.[46–49]

There are other beneficial effects of Hsp72 that could be potentially antiaging. For example, Hsp72 can have antiapoptotic effects and improve insulin resistance.[50,51] Hsp72 can also indirectly modulate inflammation as it also induces IL-10 production.[52] Furthermore, as we shall discuss subsequently, eHsp72 can exert protective actions by affecting the neuro-endocrine-immune network, particularly the interactions between cytokines and neuroendocrine agents.

Interplay among Hsp72, cytokines, the HPA axis, and the sympathetic nervous system during aging

As part of its proinflammatory effects, eHsp72 induces the production of IL-1, IL-6, and TNF-α.[16] In turn, these cytokines can stimulate the HPA axis, an effect that results in the increased release of glucocorticoids and other adrenal steroids that are powerful anti-inflammatory agents.[23] Thus, Hsp72 could trigger a feedback anti-inflammatory mechanism. Glucocorticoids not only inhibit proinflammatory cytokines synthesis and release, but also increase IL-10 production,[53,54] a prototypic anti-inflammatory cytokine. Thus, Hsp72–glucocorticoid–IL-10 interactions appear to be relevant to keep inflammaging under control.

The sympathetic nervous system (SNS), another component of the neuroendocrine–immune network that is often activated concomitantly with the HPA axis, is also functionally linked to the effects of Hsp72 on immunity. Indeed, psychophysical stress leads to increased sympathetic activity and noradrenaline (NA), the main sympathetic neurotransmitter, stimulates eHsp72 production by neutrophils.[7] This neurotransmitter also interferes with the synthesis of proinflammatory cytokines and promotes IL-10 production.[55]

The evidence discussed indicates that, in addition to direct actions, Hsp72 has the capacity to affect the production of cytokines that, in turn, induce neuroendocrine responses. Thus, Hsp72 can be considered as an intrinsic component of the immune-neuro-endocrine network. As mentioned, this network, which involves all physiological control systems that underlie growth, development, reproduction, and aging, contributes to immunoregulation, mediates neuroendocrine metabolic adjustments during disease, and maintains adaptive brain functions.

Physiological levels of both eHsp72 and corticosterone can stimulate certain aspects of the innate immune response (chemotaxis, phagocytosis, and microbicidal activity) and prevent infectious diseases, particularly under stress.[18,20,56,57] In addition, proinflammatory cytokines produced during an immune response stimulate the release of glucocorticoids that can in turn negatively feedback and inhibit the production of such cytokines.[55,58–60] A tight control of the production of proinflammatory agents and glucocorticoids levels would be required because an imbalance can contribute to pathology, for example, during sepsis.[61] Furthermore, when increased levels of glucocorticoids are reached at the peak of a strong immune response, the immune-suppressive effects of this hormone would preferentially affect noncommitted cells, leading to both an increase in the specificity of the response and to an inhibition of by-standing inflammatory processes.[23]

There is also *in vivo* and *in vitro* evidence indicating that glucocorticoids can induce the production of Hsp72,[62] and that this protein is produced in the adrenal glands when the HPA axis is stimulated during psycho/motor stress.[63] These findings indicate the existence of functional interactions between Hsp72 and the HPA axis that, because of their known links, for example, during the response to stress, would also involve the SNS.

We have recently reported that eHsp72 is detectable only in plasma of aged (13-month) and very old (26-month) mice, but not in younger counterparts (3 months).[64] Very old mice showed the highest eHsp72 plasma concentrations together with the highest corticosterone and the lowest glucose levels of the three age groups studied. Twenty-six-month-old animals also exhibited a better glucose tolerance than 13-month-old mice, which was similar to that of younger mice. Furthermore, while immune cells from very old animals produced low levels of IL-6, the capacity to produce IL-10 was much higher than in younger mice. A possible explanation to our findings could be that the presence of a danger signal such as eHsp72 (associated with infection-induced inflammation and stress situations) may serve to facilitate immune function and metabolic adjustments. Furthermore, the release of eHsp72 could be a countermeasure against deleterious effects of other inflammatory cell mediators, such as cyclooxygenase, nitric oxide, and reactive oxygen species. Indeed, circulating Hsp72 could play a protective role against deleterious conditions related to inflammation, such as infections, and related disorders during systemic psycho/physical stress conditions.[22] The evidence that enhanced oxidative stress during a maximal exercise test in aged individuals does not result in changes in the number of apoptotic or necrotic cells supports this possibility. This effect was attributed to the protective actions of the increased production of Hsp72 by lymphoid cells noticed during this stressful condition.[46,65,66] Although it is not known whether the stress-mediated protective increase of Hsp72 is also manifested in human centenarians, different factors could explain their low Hsp72 levels, while the opposite is observed in aged mice. A straightforward speculation is that humans have one of the longest lifespans of the animal kingdom and live in quite different conditions than do laboratory mice. Centenarians, as a select population exposed to multiple environmental

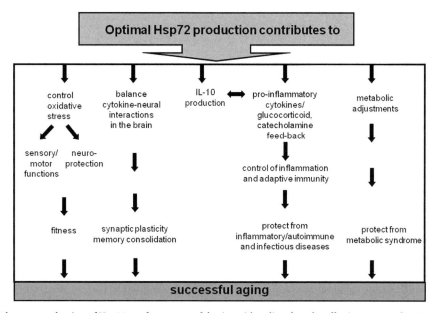

Figure 1. Adequate production of Hsp72 can favor successful aging, either directly or by affecting neuroendocrine and metabolic mechanisms. Mechanisms by which a balanced production of Hsp72 can favor successful aging are schematically represented. These proteins can interfere with excessive oxidative stress at peripheral and brain levels by supporting sensorimotor functions, exerting neuronal protection from inflammatory insults, and balancing the production of cytokines in the brain that are involved in synaptic plasticity and memory consolidation. Hsp72 can control the production of pro- and anti-inflammatory cytokines either directly or by modulating immune-mediated interactions with the HPA axis and the SNS. Well-balanced Hsp72 production can also contribute to the restriction of inflammatory processes without affecting general host defenses and, by interacting with cytokines, can mediate adjustments of metabolism that interfere with the expression of diseases linked to metabolic syndrome.

insults during prolonged life, may have developed different ways to downregulate autoinflammatory responses. The increased levels of cortisol and IL-10 that centenarians have might be enough to restrict inflammation under basal conditions, without the need to also increase Hsp72 levels. Because the lifespan of mice is much shorter than that of humans and because mice live in protected laboratory conditions, a cumulative response to inflammatory stimuli in those animals that live long may require higher Hsp72 concentrations to complement elevated corticosterone and IL-10 levels, in order to maintain an adaptive "inflammaging–anti-inflammaging" balance even under basal conditions.

Overview and proposal

Franceschi and collaborators have proposed that "the key to successful aging and longevity is to decrease chronic inflammation without compromising an acute response when exposed to pathogens."[36] Although in general we agree with this concept, we would like to add that well-balanced

immune-neuro-endocrine interactions would also be required to achieve a successful aging and longevity. In the following, some evidence in support of the notion that danger signals such as Hsp72 are relevant for an appropriate functioning of this network is discussed.

Hsp72 stimulates the production of proinflammatory cytokines that can activate the HPA axis and the SNS.[16,23] Thus, due to its capacity to affect the production of cytokines that in turn induce neuroendocrine responses, Hsp72 is an intrinsic component of the immune-neuro-endocrine network (Fig. 1). In addition to the mentioned immune-regulatory responses, this network can mediate neuroendocrine metabolic adjustments and influence adaptive brain functions. For example, endogenous IL-1 acting at peripheral and brain levels can reset glucose homeostasis. Such resetting is relevant for a redistribution of energy toward immune cells during infective and inflammatory process and to neural cells during increased neuronal activity.[67,68] Furthermore, IL-1 exerts antidiabetic effects in normal and insulin-resistant obese diabetic mice.[68–72] In contrast, this cytokine favors insulin resistance and the development of metabolic syndrome in animals subject to a high fat diet.[73,74] It is worth noting that Hsp72 stimulates insulin-induced glucose uptake and protects against obesity-induced insulin resistance. [51,75]

There is now clear evidence that cytokines, such as IL-1 and IL-6, are important to support physiologic brain functions,[76–80] and that Hsp72 has neuroprotective effects.[49,81] For example, a moderate overproduction of IL-1 in the brain during increased neuronal activity is necessary to maintain synaptic plasticity and memory consolidation.[79,80] Hsp72 also supports synaptic plasticity, and increases in its production are positively correlated with IL-1β.[82,83] It is conceivable that interactions among these proteins could affect memory consolidation.

The capacity of Hsp72 to affect and to be affected directly or indirectly by immune and neuroendocrine mechanisms indicates that this heat shock protein is part of the signaling system that can modulate the activity of the neuro-endocrine-immune network. Furthermore, during stress (both cellular and systemic), such modulatory effects of Hsp72 in association with its direct actions on cells from the different systems could influence the aging process.

Collectively considering the information discussed led to the proposal that an adequate production of Hsp72 during the course of life is relevant for successful aging. The effects that this protein could exert during aging are schematically represented in Figure 1. In summary, Hsp72 can favor the course toward successful aging:

- by modulating interactions among pro- and anti-inflammatory cytokines, glucocorticoids, and adrenergic neurotransmitters that underlie inflammaging; when such interactions are well balanced, Hsp72 would prevent or delay the tendency linked to aging to express disabling inflammatory pathologies, without affecting mechanisms of host defenses against infectious agents;
- by moderating oxi-inflammation and favoring sensory/motor activity and neuroprotection;
- by controlling cytokine-mediated adjustments of energy metabolism in order to reduce the risk of metabolic syndrome during aging; and

- by promoting cytokine–neuron interactions at brain level that are known to support cognitive and other brain functions, e.g., memory consolidation.

We propose that eHsp72 may play a key role in the maintenance of a well-balanced neuro-immune-endocrine network, as its release can be the cause or the consequence of the development of inflammatory responses and excessive oxidative processes, particularly during stressful situations. Indeed, basal blood levels of Hsp72 may not reflect the magnitude of the increase and use of this protein that occurs under stress and other danger conditions. Disruptions in the connections of the neuro-immune-endocrine network are an inexorable condition during the aging process. Mediators such as Hsp72 may serve to prolong its stabilization by fine-tuning the connectivity of the multiple components of the neuro-immune-endocrine network. Thus, both low and high circulating levels of eHsp72 in centenarian organisms may be indicative of immunophysiological adaptations that have allowed some animals of a given species to have an exceptionally long life span.

Although, as represented in Figure 1, we propose that an appropriate balance in Hsp72 production and its effects on the neuro-immune-endocrine network are relevant factors for the aging process, it is important to emphasize that the assessment of a single marker of longevity can be misleading. Indeed, the aging process is a consequence of a plethora of factors that involve physiological adaptations. Thus, Hsp72 may be included as one marker of aging once its relation with other phenotypical components of longevity is established.

Conflicts of interest

The authors declare no conflicts of interest.

References

1. Lindquist, S. & E.A. Craig. 1988. The heat-shock proteins. *Ann. Rev. Genet.* **22:** 631–677.
2. Asea, A. 2005. Stress proteins and initiation of immune response: chaperokine activity of hsp72. *Exerc. Immunol. Rev.* **11:** 34–45.
3. Fleshner, M. *et al.* 2004. Cat exposure induces both intra- and extracellular Hsp72: the role of adrenal hormones. *Psychoneuroendocrinology* **29:** 1142–1152.
4. Lindquist, S. 1986. The heat-shock response. *Annu. Rev. Biochem.* **55:** 1151–1191.
5. Asea, A. 2007. Hsp72 release: mechanisms and methodologies. *Methods* **43:** 194–198.

6. Clayton, A. *et al.* 2005. Induction of heat shock proteins in B-cell exosomes. *J. Cell Sci.* **118:** 3631–3638.

7. Giraldo, E., G. Multhoff & E. Ortega. 2010. Noradrenaline increases the expression and release of Hsp72 by human neutrophils. *Brain Behav. Immun.* **24:** 672–677.

8. Hunter-Lavin, C. *et al.* 2004. Hsp70 release from peripheral blood mononuclear cells. *Biochem. Biophys. Res. Commun.* **324:** 511–517.

9. Giraldo, E. *et al.* 2008. Influence of gender and oral contraceptives intake on innate and inflammatory response. Role of neuroendocrine factors. *Mol. Cell Biochem.* **313:** 147–153.

10. Pockley, A.G., J. Shepherd & J.M. Corton. 1998. Detection of heat shock protein 70 (Hsp70) and anti-Hsp70 antibodies in the serum of normal individuals. *Immunol. Invest.* **27:** 367–377.

11. Campisi, J., T.H. Leem & M. Fleshner. 2003. Stress-induced extracellular Hsp72 is a functionally significant danger signal to the immune system. *Cell Stress Chaperones* **8:** 272–286.

12. Pockley, A.G. *et al.* 2002. Circulating heat shock protein and heat shock protein antibody levels in established hypertension. *J. Hypertens.* **20:** 1815–1820.

13. Pockley, A.G. *et al.* 2003. Serum heat shock protein 70 levels predict the development of atherosclerosis in subjects with established hypertension. *Hypertension* **42:** 235–238.

14. Ortega, E. *et al.* 2009. The effect of stress-inducible extracellular Hsp72 on human neutrophil chemotaxis: a role during acute intense exercise. *Stress* **12:** 240–249.

15. Johnson, J.D. *et al.* 2005. Adrenergic receptors mediate stress-induced elevations in extracellular Hsp72. *J. Appl. Physiol.* **99:** 1789–1795.

16. Asea, A. *et al.* 2000. HSP70 stimulates cytokine production through a CD14-dependant pathway, demonstrating its dual role as a chaperone and cytokine. *Nat. Med.* **6:** 435–442.

17. Asea, A. *et al.* 2002. Novel signal transduction pathway utilized by extracellular HSP70: role of toll-like receptor (TLR) 2 and TLR4. *J. Biol. Chem.* **277:** 15028–15034.

18. Giraldo, E. *et al.* 2010. Exercise-induced extracellular 72 kDa heat shock protein (Hsp72) stimulates neutrophil phagocytic and fungicidal capacities via TLR-2. *Eur. J. Appl. Physiol.* **108:** 217–225.

19. Hinchado, M.D., E. Giraldo & E. Ortega. 2012. Adrenoreceptors are involved in the stimulation of neutrophils by exercise-induced circulating concentrations of Hsp72: cAMP as a potential "intracellular danger signal." *J. Cell. Physiol.* **227:** 604–608.

20. Ortega, E. *et al.* 2006. Role of Hsp72 and norepinephrine in the moderate exercise-induced stimulation of neutrophils' microbicide capacity. *Eur. J. Appl. Physiol.* **98:** 250–255.

21. Wang, R. *et al.* 2006. HSP70 enhances macrophage phagocytosis by interaction with lipid raft-associated TLR-7 and upregulating p38 MAPK and PI3K pathways. *J. Surg. Res.* **136:** 58–69.

22. Fleshner, M., J. Campisi & J.D. Johnson. 2003. Can exercise stress facilitate innate immunity? A functional role for stress-induced extracellular Hsp72. *Exer. Immunol. Rev.* **9:** 6–24.

23. Besedovsky, H.O. & A. del Rey. 2007. Physiology of psychoneuroimmunology: a personal view. *Brain Behav. Immun.* **21:** 34–44.

24. De la Fuente, M. & J. Miquel. 2009. An update of the oxidation-inflammation theory of aging: the involvement of the immune system in oxi-inflamm-aging. *Curr. Pharm. Design* **15:** 3003–3026.

25. Fabris, N. 1991. Neuroendocrine-immune interactions: a theoretical approach to aging. *Arch. Gerontol. Geriat.* **12:** 219–230.

26. Pawelec, G. 1995. Molecular and cell biological studies of ageing and their application to considerations of T lymphocyte immunosenescence. *Mech. Ageing Dev.* **79:** 1–32.

27. Pawelec, G. *et al.* 1995. Immunosenescence: ageing of the immune system. *Immunol. Today* **16:** 420–422.

28. Vellas, B.J., J.-L. Albaréde & P.J. Garry. 1993. *Facts and Research in Gerontology*, Vol. 7. J.L. Albarède, P.J. Garry & P. Vellas, Eds.: 23–29. Serdi. Paris.

29. Ortega, E. *et al.* 1992. Effect of physical activity stress on the phagocytic process of peritoneal macrophages from old guinea pigs. *Mech. Ageing Dev.* **65:** 157–165.

30. Fagiolo, U. *et al.* 1993. Increased cytokine production in mononuclear cells of healthy elderly people. *Eur. J. Immunol.* **23:** 2375–2378.

31. Ortega, E., J.J. Garcia & M. De La Fuente. 2000. Ageing modulates some aspects of the non-specific immune response of murine macrophages and lymphocytes. *Exp. Physiol.* **85:** 519–525.

32. Ortega, E. *et al.* 2000. Changes with aging in the modulation of macrophages by norepinephrine. *Mech. Ageing Dev.* **118:** 103–114.

33. Arranz, L. *et al.* 2010. Differential expression of Toll-like receptor 2 and 4 on peritoneal leukocyte populations from long-lived and non-selected old female mice. *Biogerontology* **11:** 475–482.

34. Franceschi, C. 2007. Inflammaging as a major characteristic of old people: can it be prevented or cured? *Nutr. Rev.* **65:** S173–176.

35. Franceschi, C. *et al.* 2000. Inflamm-aging. An evolutionary perspective on immunosenescence. *Ann. N.Y. Acad. Sci.* **908:** 244–254.

36. Franceschi, C. *et al.* 2007. Inflammaging and anti-inflammaging: a systemic perspective on aging and longevity emerged from studies in humans. *Mech. Ageing Dev.* **128:** 92–105.

37. Harman, D. 1956. Aging: a theory based on free radical and radiation chemistry. *J. Gerontol.* **11:** 298–300.

38. Kulinsky, V.I. 2007. Biochemical aspects of inflammation. *Biochemistry* **72:** 595–607.

39. Miquel, J. *et al.* 2009. In *Actualizaciones En Aspectos Básicos Y Clínicos Del Envejecimiento Y La Fragilidad*. L. Rodríguez, Ed.: 143–166. Novartis. Madrid.

40. Jin, X. *et al.* 2004. Serum and lymphocyte levels of heat shock protein 70 in aging: a study in the normal Chinese population. *Cell Stress Chaperones* **9:** 69–75.

41. Njemini, R. *et al.* 2011. Circulating heat shock protein 70 in health, aging and disease. *BMC Immunology* **12:** 24.

42. Njemini, R., C. Demanet & T. Mets. 2004. Inflammatory status as an important determinant of heat shock protein 70 serum concentrations during aging. *Biogerontology* **5:** 31–38.

43. Rea, I.M., S. McNerlan & A.G. Pockley. 2001. Serum heat shock protein and anti-heat shock protein antibody levels in aging. *Exp. Gerontol.* **36:** 341–352.

44. Terry, D.F. *et al.* 2004. Cardiovascular disease delay in centenarian offspring: role of heat shock proteins. *Ann. N.Y. Acad. Sci.* **1019:** 502–505.

45. Terry, D.F. *et al.* 2006. Serum heat shock protein 70 level as a biomarker of exceptional longevity. *Mech. Ageing Dev.* **127:** 862–868.

46. Chiu, H.Y., L.Y. Tsao & R.C. Yang. 2009. Heat-shock response protects peripheral blood mononuclear cells (PBMCs) from hydrogen peroxide-induced mitochondrial disturbance. *Cell Stress Chaperones* **14:** 207–217.

47. Tang, D. *et al.* 2007. Nuclear heat shock protein 72 as a negative regulator of oxidative stress (hydrogen peroxide)-induced HMGB1 cytoplasmic translocation and release. *J. Immunol.* **178:** 7376–7384.

48. Volloch, V. & S. Rits. 1999. A natural extracellular factor that induces Hsp72, inhibits apoptosis, and restores stress resistance in aged human cells. *Exp. Cell Res.* **253:** 483–492.

49. Xu, L. *et al.* 2010. Astrocyte targeted overexpression of Hsp72 or SOD2 reduces neuronal vulnerability to forebrain ischemia. *Glia* **58:** 1042–1049.

50. Gupte, A.A., G.L. Bomhoff & P.C. Geiger. 2008. Age-related differences in skeletal muscle insulin signaling: the role of stress kinases and heat shock proteins. *J. Appl. Physiol.* **105:** 839–848.

51. Gupte, A.A. *et al.* 2011. Acute heat treatment improves insulin-stimulated glucose uptake in aged skeletal muscle. *J. Appl. Physiol.* **110:** 451–457.

52. Luo, X. *et al.* 2008. Release of heat shock protein 70 and the effects of extracellular heat shock protein 70 on the production of IL-10 in fibroblast-like synoviocytes. *Cell Stress Chaperones* **13:** 365–373.

53. Mozo, L., A. Suarez & C. Gutierrez. 2004. Glucocorticoids up-regulate constitutive interleukin-10 production by human monocytes. *Clin. Exp. Allergy* **34:** 406–412.

54. Verhoef, C.M. *et al.* 1999. The immune suppressive effect of dexamethasone in rheumatoid arthritis is accompanied by upregulation of interleukin 10 and by differential changes in interferon gamma and interleukin 4 production. *Ann. Rheum. Dis.* **58:** 49–54.

55. Elenkov, I.J. & G.P. Chrousos. 2002. Stress hormones, proinflammatory and antiinflammatory cytokines, and autoimmunity. *Ann. N.Y. Acad. Sci.* **966:** 290–303.

56. Forner, M.A. *et al.* 1995. A study of the role of corticosterone as a mediator in exercise-induced stimulation of murine macrophage phagocytosis. *J. Physiol.* **488:** 789–794.

57. Ortega, E., M.A. Forner & C. Barriga. 1997. Exercise-induced stimulation of murine macrophage chemotaxis: role of corticosterone and prolactin as mediators. *J. Physiol.* **498:** 729–734.

58. Besedovsky, H. *et al.* 1986. Immunoregulatory feedback between interleukin-1 and glucocorticoid hormones. *Science* **233:** 652–654.

59. Besedovsky, H.O. *et al.* 1991. Cytokines as modulators of the hypothalamus-pituitary-adrenal axis. *J. Steroid Biochem. Mol. Biol.* **40:** 613–618.

60. del Rey, A. *et al.* 1987. Interleukin-1 and glucocorticoid hormones integrate an immunoregulatory feedback circuit. *Ann. N.Y. Acad. Sci.* **496:** 85–90.

61. Besedovsky, H.O. & A. del Rey. 2006. Regulating inflammation by glucocorticoids. *Nat. Immunol.* **7:** 537.

62. Sun, L. *et al.* 2000. Activation of HSF and selective increase in heat-shock proteins by acute dexamethasone treatment. *Am. J. Physiol. Heart Circ. Physiol.* **278:** H1091–H1097.

63. Campisi, J. *et al.* 2003. Habitual physical activity facilitates stress-induced HSP72 induction in brain, peripheral, and immune tissues. *Am. J. Physiol. Regul. Integr. Comp. Physiol.* **284:** R520–R530.

64. Ortega, E. *et al.* 2011. Altered Hsp72-induced release of inflammatory cytokines and circulating levels of Hsp72, glucose, and corticosterone during aging. *Neuroimmunomodulation* **18:** 395.

65. Murlasits, Z. *et al.* 2006. Resistance training increases heat shock protein levels in skeletal muscle of young and old rats. *Exp. Gerontol.* **41:** 398–406.

66. Simar, D. *et al.* 2004. Effect of age on Hsp72 expression in leukocytes of healthy active people. *Exp. Gerontol.* **39:** 1467–1474.

67. Besedovsky, H.O. & A. del Rey. 2011. Central and peripheral cytokines mediate immune-brain connectivity. *Neurochem. Res.* **36:** 1–6.

68. del Rey, A. *et al.* 2006. IL-1 resets glucose homeostasis at central levels. *Proc. Natl. Acad. Sci. U.S.A.* **103:** 16039–16044.

69. Besedovsky, H.O. & A. del Rey. 2010. Interleukin-1 resets glucose homeostasis at central and peripheral levels: relevance for immunoregulation. *Neuroimmunomodulation* **17:** 139–141.

70. del Rey, A. & H. Besedovsky. 1989. Antidiabetic effects of interleukin 1. *Proc. Natl. Acad. Sci. U.S.A.* **86:** 5943–5947.

71. Matsuki, T. *et al.* 2003. IL-1 plays an important role in lipid metabolism by regulating insulin levels under physiological conditions. *J. Exp. Med.* **198:** 877–888.

72. Somm, E. *et al.* 2005. Decreased fat mass in interleukin-1 receptor antagonist-deficient mice: impact on adipogenesis, food intake, and energy expenditure. *Diabetes* **54:** 3503–3509.

73. de Roos, B. *et al.* 2009. Attenuation of inflammation and cellular stress-related pathways maintains insulin sensitivity in obese type I interleukin-1 receptor knockout mice on a high-fat diet. *Proteomics* **9:** 3244–3256.

74. Perrier, S., F. Darakhshan & E. Hajduch. 2006. IL-1 receptor antagonist in metabolic diseases: Dr Jekyll or Mr Hyde? *FEBS Lett.* **580:** 6289–6294.

75. Chung, J. *et al.* 2008. HSP72 protects against obesity-induced insulin resistance. *Proc. Natl. Acad. Sci. U.S.A.* **105:** 1739–1744.

76. Balschun, D. *et al.* 2003. Hippocampal interleukin-1 beta gene expression during long-term potentiation decays with age. *Ann. N.Y. Acad. Sci.* **992:** 1–8.

77. Balschun, D. *et al.* 2004. Interleukin-6: a cytokine to forget. *FASEB J.* **18:** 1788–1790.

78. Goshen, I. *et al.* 2009. Environmental enrichment restores memory functioning in mice with impaired IL-1 signaling via reinstatement of long-term potentiation and spine size enlargement. *J. Neurosci.* **29:** 3395–3403.

79. Goshen, I. *et al.* 2007. A dual role for interleukin-1 in hippocampal-dependent memory processes. *Psychoneuroendocrinology* **32:** 1106–1115.

80. Schneider, H. *et al.* 1998. A neuromodulatory role of interleukin-1beta in the hippocampus. *Proc. Natl. Acad. Sci. U.S.A.* **95:** 7778–7783.

81. Giffard, R.G. & M.A. Yenari. 2004. Many mechanisms for hsp70 protection from cerebral ischemia. *J. Neurosurg. Anesthesiol.* **16:** 53–61.

82. Lin, Y.W. *et al.* 2004. Heat-shock pretreatment prevents suppression of long-term potentiation induced by scopolamine in rat hippocampal CA1 synapses. *Brain Res.* **999:** 222–226.

83. Nickerson, M. *et al.* 2005. Physical activity alters the brain Hsp72 and IL-1beta responses to peripheral E. coli challenge. *Am. J. Physiol. Regul. Integr. Comp. Physiol.* **289:** R1665–R1674.

Ann. N.Y. Acad. Sci. ISSN 0077-8923

ANNALS OF THE NEW YORK ACADEMY OF SCIENCES
Issue: *Neuroimmunomodulation in Health and Disease*

Regulation of intestinal morphology and GALT by pituitary hormones in the rat

Luz María Cárdenas-Jaramillo,[1] Andrés Quintanar-Stephano,[2] Rosa Adriana Jarillo-Luna,[1] Víctor Rivera-Aguilar,[3] Gabriela Oliver-Aguillón,[1] Rafael Campos-Rodríguez,[4] Kalman Kovacs,[5] and Istvan Berczi[2,6]

[1]Departamento de Morfología, Laboratorio de Morfología Sección de Estudios de Posgrado e Investigación, Escuela Superior de Medicina, IPN, México. [2]Departamento de Fisiología y Farmacología Centro de Ciencias Básicas, Universidad Autónoma de Aguascalientes, Aguascalientes, México. [3]Departamento de Microbiología, UBIPRO, FES-Iztacala, UNAM, México. [4]Departamento de Bioquímica, Escuela Superior de Medicina, IPN, México. [5]Department of Medicine Pathology, St. Michael's Hospital, Toronto, ON, Canada. [6]Department of Immunology, Faculty of Medicine, University of Manitoba, Winnipeg, MB, Canada

Address for correspondence: Dr. Andrés Quintanar-Stephano, Departamento de Fisiología y Farmacología, Centro de Ciencias Básicas, Universidad Autónoma de Aguascalientes, Av. Universidad # 940, Col. Ciudad Universitaria, Aguascalientes, Ags. CP 20131, México. aquinta@correo.uaa.mx

Here, the effects of neurointermediate (NIL), anterior (AL), and total hypophysectomy (HYPOX) on ileal mucosa cells and gut-associated lymphoid tissue (GALT) are reported. Compared with the sham-operated (SHAM) rats, the villi height and goblet cells numbers were significantly decreased in all groups. Lamina propria area decreased in AL and HYPOX, but not in NIL animals. CD8+ but not CD4+ lymphocytes decreased in the HYPOX and NIL groups. Paneth cells did not change, while IgA cells, IgM cells, and secretory IgA were significantly decreased in all groups. NIL but not AL animals lost significant numbers of IgA cells and secretory IgA. In summary, pituitary hormones exert lobe-specific regulatory effects on the gut and on GALT.

Keywords: hypophysectomy; anterior and neurointermediate pituitary lobectomy; pituitary hormones; gut morphology; gut associated lymphoid tissue

Introduction

Pituitary hormones regulate growth, metabolism, reproduction, sexual differentiation, and maturation, aging, immunological, and pathophysiological defense mechanisms of the body. Thus, the pituitary gland still functions as the "conductor of the endocrine orchestra," which regulates the entire animal/human organism.[1]

Hans Selye pointed out first that the hypothalamus–pituitary–adrenal (HPA) axis is activated by diverse "nocuous" agents that act on the central nervous system (CNS) first, and then through the HPA axis. Adrenal glucocorticoids (GCs) cause atrophy in the thymus gland, spleen, and other lymphoid organs.[2,3] Selye called this response the *stress response*, and eventually he concluded that *stressors*

induce a defense reaction that he called the *general adaptation syndrome*.[4] Today we know a great deal about the HPA axis and its role in immunoregulation; it suppresses adaptive immunity (ADIM) and promotes innate immunity (INIM) through moderating the excess of cytokine production, and it has an anti-inflammatory effect in general. Without the HPA axis excessive cytokine production occurs during acute illness, and high level of cytokines can kill the host.[5] The HPA axis also regulates numerous physiological and pathophysiological reactions in the body. GCs, for example, are required for normal lymphocyte development and function.[6]

In the paraventricular nucleus (PVN) of the hypothalamus, corticotropin-releasing hormone (CRH) is secreted and regulates the release of adrenocorticotropic hormone (ACTH) from the

doi: 10.1111/j.1749-6632.2012.06648.x

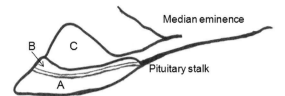

Figure 1. Sagital view of the rat pituitary gland showing the several lobes, which were selectively removed: adenohypophyseal lobe (A), neurointermediate pituitary lobe (A + B), and total hypophysectomy (A + B + C lobes).

pituitary gland, which in turn stimulates the release of GC from the adrenal cortex. Vasopressin (VP) supports the stimulatory effects of CRH on the HPA axis during acute illness. However, in chronic inflammatory disease and during homeostatic conditions, VP is the major regulator of the HPA axis. VP also regulates recovery and healing of the host organism from disease.[5]

The growth and lactogenic hormone (GLH) family includes growth hormone (GH), prolactin (PRL), and placental lactogens (PL). Insulin-like growth factor-I (IGF-I) is also involved in the growth promoting effect of GLHs.[7] GLH family members share their signal transduction pathway (JAK-STAT) with type I cytokines, which are important immunoregulators. Consequently, GLHs share an immunoregulatory function with type I cytokines.[8] GLHs promote body growth *in utero* during fetal development and in young animals and humans. Lymphocyte growth and differentiation of the ADIM system also relies on GLHs. Lymphocyte clones of the ADIM system must proliferate in order to produce an army of effector cells, which, after differentiation, defend the host against specific pathogens. GLH is capable of maintaining adaptive immunocompetence.[5] Memory cells are capable of functioning independently of GLH and survive major disasters and HYPOX as well.[9]

Oxytocin is an immunosuppressor.[10] Dopamine suppresses PRL secretion and suppresses ADIM; it induces IL-6 and Th17 cells, promotes inflammatory disease, and induces neurotoxicity.[11–14] Thyroxin (T4) has a stimulatory effect on GH and PRL, and for this reason T4 has an immunostimulatory effect (unpublished data). Somatostatin/cortistatin have been recently recognized as inhibitory regulators of immune function.[15] Ghrelin is the endogenous ligand for the GH-secretagogue receptor, and

consequently is expected to act as an immunostimulator.[16] Recent work has shown that CRH has two receptors that should enrich the regulatory effect of CRH. Urocortins are CRH analogues, and annexins are glucocorticoid regulated inflammatory mediators.[18,19] These new regulators demonstrate that pituitary immunoregulation is under multifactorial control. Numerous other hormones have been coined as *immunomodulators,* which means that, while they are not able to induce or suppress immune reactions, once an immune reaction is already ongoing, they are capable of modulating the response.[20] The effect of pituitary hormones on intestinal mucosa and gut-associated lymphoid tissue (GALT) is poorly understood. Here, we present histological evidence for the effects of pituitary hormone lobe-specific deficiency on both ileum mucosa and GALT.

Methods

Three-month-old male Wistar rats were divided in the following groups: (1) sham-operated (SHAM), (2) neurointermediate pituitary lobectomized (NIL), (3) anterior pituitary lobectomized (AL), and (4) total hypophysectomized (HYPOX).

Surgeries

Rats were anesthetized with vapors of Sevorane and the tracheas were cannulated by mouth. Atropin (0.06 mg/100 g body weight) was administered subcutaneously (sc) to each rat in order to inhibit excess mucus production. SHAM, NIL, AL, and HYPOX were performed under a dissecting microscope through the parapharyngeal-transsphenoidal approach (Fig. 1; AL = lobe A resected, NIL = B and C lobes resected, and HYPOX = A, B, and C lobes resected). In SHAM animals, the operation was terminated when the pituitary capsule was surgically opened and the pituitary was visualized. For AL, the adenohypophysis was divided in two halves and then aspirated. For NIL, the adenohypophysis was lifted, and under direct view the neurointermediate lobe was aspirated with a special curved needle. For HYPOX, the entire hypophysis was removed.

Ileum secretory IgA

Animals were sacrificed 4 weeks after surgeries; ilea were dissected and washed with 5 mL of washing fluid (containing NaCl, 8 g; KCl, 0.2 g; Na_2HPO_4,

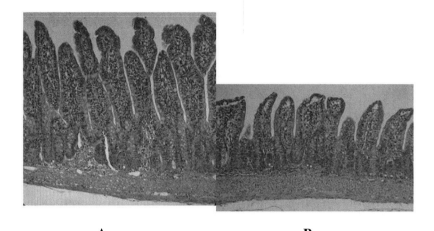

Figure 2. Microphotographs of the ileum mucosa from SHAM (A) and HYPOX (B) animals. Note the decreased villi height of the HYPOX animal and the similar submucosa width. Stain H-E. Magnification: 200x.

1.15 g; KH_2PO_4, 0.2 g; distilled water, 100 mL). Secretory IgA (μg/mL) was measured in the washing fluid by ELISA.[21]

Histology

Ilea were fixed in 10% neutral formalin and processed in paraffin for histology. Histological slides were stained with hematoxylin-eosin (H-E) and periodic acid shift-H (PAS-H), or immunostained for $CD4^+$ and $CD8^+$ lymphocytes, Paneth cells, and IgA^+ and IgM^+ cells.

Height of intestinal villi, lamina propria area, goblet cell number per villi, Paneth, IgA, and IgM cells numbers, and lamina propria and intraepithelial $CD4^+$ and $CD8^+$ cells were counted per area unit (CD4, CD8, IgA, and IgM antibodies for immunohistochemistry were acquired from Bio SB Inc., Santa Barbara, CA).

Results

As compared with the SHAM, the height of ilea villi decreased significantly in AL, NIL, and HYPOX groups (25%, 22%, and 34%, respectively, $P < 0.001$, Figs. 2A, 2B and 3A), whereas the lamina propria area decreased significantly in AL (38%) and HYPOX (27%) groups (Fig. 3B). Compared with the SHAM, goblet cell numbers decreased significantly in NIL, AL, and HYPOX groups (Fig. 3B).

Compared with the SHAM group lamina propria $CD4^+$ lymphoctyes, nonsignificant differences with the remaining groups occurred (Fig. 4A); however, AL animals showed a significantly higher number of

$CD4^+$ lymphoctyes compared with the NIL group. Compared with the SHAM and AL groups, lamina propria $CD8^+$ lymphocytes were significantly fewer in the HYPOX group (Fig. 4A), whereas no differences between the HYPOX and NIL groups were apparent. Intraepithelial $CD4^+$ lymphocytes were present in low numbers in all experimental groups, indicating that surgical manipulation of the animals had no effect on this parameter (Fig. 4B). Observe in the same figure the similar numbers of intraepithelial $CD8^+$ lymphocytes in SHAM, AL, and HYPOX groups, whereas a significant diminution occurred in the NIL animals ($P < 0.001$ versus SHAM, AL, and HYPOX).

Table 1 shows that surgical modification of pituitary hormones did not significantly affect Paneth cells in the several experimental groups, whereas when compared with the SHAM group, IgM and IgA cells significantly decreased in NIL ($P < 0.001$), AL, and HYPOX ($P < 0.01$) animals. Intestinal IgA secretion similarly decreased by NIL ($P < 0.001$), AL, and HYPOX ($P < 0.01$) (Table 1).

Discussion

Previous experiments in rats indicated that NIL protects against experimental autoimmune encephalitis (EAE) and treatment with the vasopressin agonist drug, desmopressin, restored EAE in NIL rats.[22]

In the present work, the results indicate that the maintenance of villi height and goblet cells was dependent on anterior and posterior pituitary hormones (Figs. 3A and 3C), whereas the area of

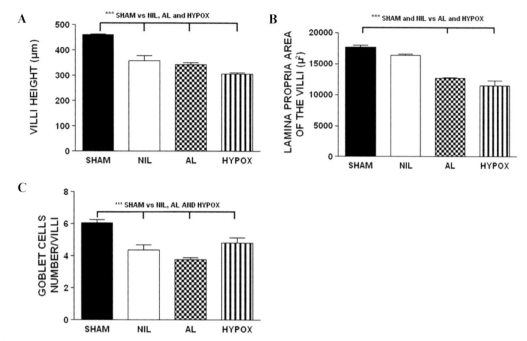

Figure 3. Effects of sham operation (SHAM), neurointermediate pituitary lobectomy (NIL), anterior lobectomy (AL), and hypophysectomy (HYPOX) on ilial villi height (μm) (A) and goblet cells number/villi (B). $n = 3$ animals/group; ***$P < 0.001$.

the lamina propria was more dependent on anterior pituitary hormones (GH, PRL, and/or the HPA axis) (Fig. 3B). These results support the view that GH play an important role in the homeostatic maintenance of the intestinal epithelium.[23] It seems that Paneth cells number was not dependent from the pituitary hormones. In GALT, CD4[+] lymphocytes were reduced in the lamina propria of NIL animals when compared with the AL group, and CD8[+] lymphocytes were decreased in NIL and HYPOX, but not in the AL group. This suggests that hormones from the posterior lobe participate in the regulation of CD4[+] and CD8[+] lymphocytes in the lamina propria (Fig. 4A). Intraepithelial CD4[+] lymphocytes were not affected by the surgical modification of pituitary hormones, whereas for intraepithelial CD8[+] lymphocytes, posterior pituitary hormones were required, as suggested by the decreased number of these cells in NIL animals and normal number numbers of cells in AL animals (Fig. 4B). The decreased number of IgM and IgA cells, and the decreased secretion of intestinal IgA in NIL, AL, and HYPOX groups, indicate that all pituitary hormones are required for homeostatic maintenance. However, the signif-

icant diminution of IgM cells and of intestinal IgA secretion, compared with the remaining groups, suggest that posterior hormones play a more important role in the maintenance of these cells (Table 1). Thus, lamina propria CD4[+] and CD8[+] lymphocytes, intraepithelial CD8[+] lymphocytes of GALT, as well as IgM cells and secretory IgA, are more dependent on posterior pituitary hormones, most likely on VP.

It is interesting to note that CD4[+] and CD8[+] T lymphocytes were no different in the SHAM animals and in the HYPOX group, which have severely damaged endocrine systems. We also observed this phenomenon with spleen cytokine levels, in which SHAM and nonoperated control animals and HYPOX rats responded similarly to challenge with adjuvant arthritis. With these observations, we proposed that the immune system is regulated by superimposed regulatory circuits that can function as part of the neuroimmune "supersystem"— but also are capable of functioning independently on their own. In HYPOX animals, the circuits of the brain and of hypophysis had been removed; therefore, homeostasis of gut cells must be maintained by neuropeptides, cytokines, chemokines, and adhesion molecules.[9] The previously described

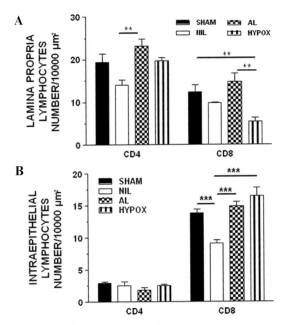

Figure 4. Effects of sham operation (SHAM), neurointermediate pituitary lobectomy (NIL), anterior lobectomy (AL), and hypophysectomy (HYPOX) on the ileal number of lamina propria (A) and intraepithelial (B) CD4[+] and CD8[+] lymphocytes/10,000 μm^2. $n = 3$ animals/group; ** $P < 0.01$ in lamina propria, and *** $P < 0.001$ in intraepithelial lymphocytes. Nonsignificant differences between CD4[+] lymphocytes were observed in the several groups.

observation is compatible with this hypothesis. These results indicate new regulators of lymphocyte recirculation and homing.

GCs have been known for some time to increase the recirculation of leukocytes, while tissue localization was believed to be mediated by adhesion molecules and cytokines.[24] AL removes GH, PRL, and destroys the HPA axis so that GCs are not produced. Therefore, further experiments are required

to determine the exact roles of GH, PRL, and GCs in gut localization of CD8[+] lymphocytes and in the development of goblet cells. AL leaves VP intact and fully functional; therefore, the conclusion that decreased counts of lamina propria CD4[+] lymphocytes and intraepithelial CD8[+] lymphocytes rely on vasopressin is realistic, since these cells decreased only in NIL animals, not in HYPOX or AL animals. It is interesting to note that CD8[+] lymphocytes depend on anterior pituitary hormones in the lamina propria, yet in the epithelium, CD8[+] lymphocytes are significantly affected by NIL, which suggests regulation by VP and not by other pituitary hormones. Perhaps T cells in the lamina propria are immature, whereas only mature cells would recirculate and extravasate into the epithelium? The indications are that different hormones regulate immature and mature CD8[+] lymphocytes, and thus T cells may have a change of regulator from anterior pituitary hormones to VP after maturation.

The *acute phase reaction* (APR)—or acute illness—is characterized by the activation of the HPA axis and leukocytosis. In the early phase of acute illness, GH and PRL will rise and there is excessive extravasation of leukocytes, which leads to relative leukopenia. At this time the ADIM system mounts a rapid peripheral defense reaction.[25] If this reaction fails, GH and PRL will be suppressed by excessive activation of the HPA axis and by stimulation of suppressor/regulatory T lymphocytes by GC and catecholamines. However, increased production of leukocytes in the bone marrow and leukocytosis will ultimately prevail. Bone marrow function is actually amplified during APR, most likely by cytokines, such as interleukin-6 (IL-6) and granulocyte-macrophage colony stimulating factor (GM-CSF).[5]

Table 1. Effects of SHAM, NIL, AL, and HYPOX on ileal Paneth, IgM, and IgA cell numbers and on intestinal IgA secretion

Group	Paneth cells	IgM cells	IgA cells	IgA, $\mu g/ml$
SHAM	2.6 ± 0.2^a	18.5 ± 1.1^a	21.9 ± 2.0^a	38.6 ± 0.1^a
NIL	2.2 ± 0.1^a	9.2 ± 0.4^b	11.2 ± 0.7^b	23.7 ± 0.5^b
AL	2.0 ± 0.05^a	13.6 ± 0.08^c	15.3 ± 1.0^b	31.8 ± 0.5^c
HYPOX	2.2 ± 0.05^a	$12.1 \pm 0.7^{b,c}$	14.7 ± 1.2^b	32.5 ± 0.8^c

NOTE: Columns with the same letter mean no statistical differences, whereas different letters mean significant differences among groups. Values are the means $\pm$ SEM.

There is evidence that GH and PRL regulate bone marrow function.[26] PRL was shown to support bone marrow transplantation and was beneficial in treatment of experimental radiation disease, burn injury, and other debilitating conditions when immune-derived type I cytokines are in short supply.[27] GLHs share the JAK/STAT signal transduction pathway with type I cytokines,[8] and therefore these hormones can substitute for the function of these cytokines. Apparently, this mechanism is responsible for the resistance of bone marrow to glucocorticoids and for increased activity during APR.

Intestinal villi and goblet cells are different elements of intestinal tissue. One could assume that GLHs, and even the HPA axis, would affect these tissues/cells. The indication from our experiments, that VP would directly affect the homing of CD4[+] T cells to the intestine, is a new finding. VP remains active during APR, but GLHs are suppressed. In other experiments, we have shown that VP stimulates cytokine production from both the adaptive and innate immune systems.[20] Is it possible that VP also stimulates intestinal tissue cytokine production, which is required for growth? Future experiments will hopefully clarify this question.

Conclusions

The results presented in this paper indicate that hormones of both the anterior and posterior pituitary lobes are required for the maintenance of lymphocyte homing to the gut, and in all probability, for survival in this tissue. GH, PRL, and the HPA axis play important roles in the maintenance of lamina propria. Vasopressin regulates the CD4[+] and CD8[+] lamina propria lymphocytes, IgM and IgA cells, and IgA secretion, as well as intraepithelial CD8[+] T cells.

Acknowledgments

We thank Alejandro Organista-Esparza, Adriana Rodríguez-Peralta, Manuel Tinajero-Ruelas, and Raul Ponce-Gallegos for technical assistance. Supported by Grants UAA PIBB-11-5; CONACYT 2011-2013; SIP-IPN-20110764; and CONACYT 80310 México.

Conflicts of interest

The authors declare no conflicts of interest.

References

1. Guyton, C.A. & J.E. Hall. 2011. Pituitary Hormones and their control by the Hypothalamus. In *Guyton and Hall Textbook of Medical Physiology*, 12th Ed.: 895–906. Saunders Elsevier. Philadelphia.
2. Selye, H. 1936. A syndrome produced by diverse nocuous agents. *Nature* **138**: 32.
3. Selye, H. 1936. Thymus and adrenals in the response of the organism to injuries and intoxication. *Br. J. Exp. Path.* **17**: 234–248.
4. Selye, H. 1946. The general adaptation syndrome and the diseases of adaptation. *J. Clin. Endocrinol.* **6**: 117–230.
5. Berczi, I., A. Quintanar-Stephano & K. Kovacs. 2009. Neuroimmune regulation in immunocompetence, acute illness, and healing. *Ann. N.Y. Acad. Sci.* **1153**: 220–239.
6. Guire, P.M., M.P. Yaeger & A. Munck. 2008. Glucocorticoid effects on the immune system. In *The Hypothalamus Pituitary Adrenal Axis*. A. del Rey, G. Chrousos & H. Besedovky, Eds. I. Berczi & A. Szentivanyi, Series Eds. Neuroimmune Biology Ed.: 147–167. Elsevier. Amsterdam.
7. Berczi, I. & A. Szentivanyi. 2003. Growth and lactogenic hormones, insulin-like growth factor and insulin. In *The Immune-Neuroendocrine Circuitry. History and Progress*. I. Berczi & A. Szentivanyi, Eds.: 129–153. Elsevier. Amsterdam.
8. Redelman, D., L.A. Welniak, D. Taub & W.J. Murphy. 2008. Neuroendocrine hormones such as growth hormone and prolactin are integral members of the immunological cytokine network. *Cell Immunol.* **252**: 111–121.
9. Berczi, I., A. Quintanar-Stephano & K. Kovacs. 2010. The brave new world of neuroimmune biology. In *New Insights to Neurimmune Biology*. I. Berczi, Ed.: 4–30. Elsevier. Amsterdam.
10. Clodi, M., G. Vila, R. Geyeregger, *et al.* 2008. Oxytocin alleviates the neuroendocrine and cytokine response to bacterial endotoxin in healthy men. *Am. J. Physiol. Endocrinol. Metab.* **295**: E686–E691.
11. Nagy, E., I. Berczi, G.E. Wren, *et al.* 1983. Immunomodulation by bromocriptine. *Immunopharmacology* **6**: 231–243.
12. Nakagome, K., M. Imamura, H. Okada, *et al.* 2011. Dopamine D1-like receptor antagonist attenuates Th17-mediated immune response and ovalbumin antigen-induced neutrophilic airway inflammation. *J. Immunol.* **186**: 5975–5982.
13. Nakano, K., K. Yamaoka, K. Hanami, *et al.* 2011. Dopamine induces IL-6-dependent IL-17 production via D1-like receptor on CD4 naive T cells and D1-like receptor antagonist SCH-23390 inhibits cartilage destruction in a human rheumatoid arthritis/SCID mouse chimera model. *J. Immunol.* **186**: 3745–3752.
14. Reynolds, A.D., D.K. Stone, J.A. Hutter, *et al.* 2010. Regulatory T cells attenuate Th17 cell-mediated nigrostriatal dopaminergic neurodegeneration in a model of Parkinson's disease. *J. Immunol.* **184**: 2261–2271
15. Baatar, D., K. Patel & D.D. Taub. 2011. The effects of ghrelin on inflammation and the immune system. *Mol. Cell. Endocrinol.* **340**: 44–58.
16. Ferone, D., M. Boschetti, E. Resmini, *et al.* 2006. Neuroendocrine-immune interactions: the role of

cortistatin/somatostatin system. *Ann. N.Y. Acad. Sci.* **1069:** 129–144.

17. Yang, L.Z., P. Tovote, M. Rayner, *et al.* 2010. Corticotropin-releasing factor receptors and urocortins, links between the brain and the heart. *Eur. J. Pharmacol.* **632:** 1–6.

18. Perretti, M. & J. Dalli. 2009. Exploiting the Annexin A1 pathway for the development of novel anti-inflammatory therapeutics. *Br. J. Pharmacol.* **158:** 936–946.

19. Beczi, I. 1986. Immunoregulation by the pituitary gland. In *Pituitary Function and Immunity*. I. Berczi, Ed.: 227–240. CRC Press. Boca Raton, FL.

20. Quintanar-Stephano, A., C. Villalobos, K. Arroyo, *et al.* 2010. Immunoregulation by the neurohypophysis. *FASEB J.* **24:** 627.1.

21. Campos-Rodríguez, R., A. Quintanar-Stephano, R.A. Jarillo-Luna, *et al.* 2006. Hypophysectomy and neurointermediate pituitary lobectomy reduce serum immunoglobulin M (IgM) and IgG and intestinal IgA responses to Salmonella enterica serovar Typhimurium infection in rats. *Infect. Immun.* **74:** 1883–1889.

22. Quintanar-Stephano, A., A. Organista-Esparza, R. Chavira-Ramírez, *et al.* 2012. Effects of neurointermediate pituitary lobectomy and desmopressin on acute experimental autoimmune encephalomyelitis in Lewis rats. *Neuroimmunomodulation* **19:** 148–157.

23. Zhang, X., J. Li & N. Li. 2002. Growth hormone improves graft mucosal structure and recipient protein metabolism in rat small bowel transplantation. *Chin. Med. J.* **115:** 732–735.

24. Salmi, M. & S. Jalkanen. 2005. Lymphocyte homing to the gut: attraction, adhesion, and commitment. *Immunol. Rev.* **206:** 100–113.

25. Dhahbar, F. 2008. Enhancing versus suppressive effect of stress on immune function: implications for immunoprotection and immunopathology. In *The Hypothalamus Pituitary Adrenal Axis*. A. del Ray, G. Chrousos & H. Besedovsky, Eds, I. Berczi & A. Szentivanyi, Series Eds. Neuroimmune Biology, Eds.: 207–224. Elsevier. Amsterdam.

26. Nagy, E. & I. Berczi. 1989. Pituitary dependence of bone marrow function. *Br. J. Haematol.* **71:** 457–462.

27. Berczi, I., R. Laatikainen & J. Pulkkinen. 2010. Sex hormones and their analogues in neuroimmune biology. *Immunol. Endo. Metabol. Agen. Medi. Chem.* **10:** 142–181.

Ann. N.Y. Acad. Sci. ISSN 0077-8923

Transforming growth factor-beta inhibits the expression of clock genes

Heidemarie Gast,[1,a] Sonja Gordic,[2,a] Saskia Petrzilka,[2] Martin Lopez,[2] Andreas Müller,[2] Anton Gietl,[3] Christoph Hock,[3] Thomas Birchler,[2,b] and Adriano Fontana[2,b]

[1]Department of Neurology, Inselspital, University Hospital Berne, University of Berne, Berne, Switzerland. [2]Institute of Experimental Immunology. [3]Division of Psychiatry Research, University of Zurich, Zurich, Switzerland

Address for correspondence: Dr. Adriano Fontana, Institute of Experimental Immunology, University of Zürich, Winterthurerstrasse 190, CH-8057 Zürich, Switzerland. adriano.fontana@usz.ch

Disturbances of sleep–wake rhythms are an important problem in Alzheimer's disease (AD). Circadian rhythms are regulated by clock genes. Transforming growth factor-beta (TGF-β) is overexpressed in neurons in AD and is the only cytokine that is increased in cerebrospinal fluid (CSF). Our data show that TGF-β2 inhibits the expression of the clock genes *Period* (*Per*)*1*, *Per2*, and *Rev-erbα*, and of the clock-controlled genes D-site albumin promoter binding protein (*Dbp)* and thyrotroph embryonic factor (*Tef*). However, our results showed that TGF-β2 did not alter the expression of brain and muscle Arnt-like protein-1 (*Bmal1*). The concentrations of TGF-β2 in the CSF of 2 of 16 AD patients and of 1 of 7 patients with mild cognitive impairment were in the dose range required to suppress the expression of clock genes. TGF-β2–induced dysregulation of clock genes may alter neuronal pathways, which may be causally related to abnormal sleep–wake rhythms in AD patients.

Keywords: sleep genes; memory; sleep deprivation; E-box

Introduction

The neuropathological hallmarks of Alzheimer's disease (AD) are neuronal loss, senile plaques, and neurofibrillary tangles in cortical and limbic regions. A major component of senile plaques is the β-amyloid (Aβ), which is cleaved by β- and γ-secretases from the amyloid precursor protein (APP). Generation and clearance of Aβ have been shown to be modulated by the immune system. Transforming growth factor-beta (TGF-β)1 drives the expression of APP in astrocytes and enhances the generation of Aβ.[1] Overexpression of TGF-β1 in the brain promotes brain inflammation, paralleled by increased deposition of brain vascular Aβ and reduction of Aβ deposits in the parenchyma.[2,3]

[a]These authors have contributed equally to this work as first authors.
[b]These authors have contributed equally to this work as last authors.

However, inhibition of TGF-β signaling in peripheral macrophages has been found to lead to an influx of macrophages into the brain and enhancement of Aβ clearance.[4] TGF-β2 induces neuronal cell death by binding to the extracellular domain of APP.[5,6] The expression of TGF-β2 is increased in glial cells located near the senile plaques in transgenic mice that overexpress the Swedish type of familial AD.[6] Furthermore, TGF-β2 is elevated in neurons of the hippocampus and cortex of patients with AD.[7] TGF-β-mediated gene transcription involves phosphorylation of Smad2 and Smad3, which thereby associate with Smad4 and translocate to the nucleus. In neurons of AD patients, phosphorylated Smad2 is overexpressed in the cytoplasm, but not in the nucleus of neurons located in neurofibrillary tangles and granulovascuolar degeneration.[8] The expression of TGF-β and pSMAD3, the activated form of the TGF-β receptor–mediated transcriptional modulator, follows a circadian pattern.[9] A recent metaanalysis of cytokine expression in AD revealed that many proinflammatory cytokines, including

doi: 10.1111/j.1749-6632.2012.06640.x
Ann. N.Y. Acad. Sci. 1261 (2012) 79–87 © 2012 New York Academy of Sciences.

IL-1β, IL-6, IL-12, IL-18, and TNF-α, are increased in the blood of patients. However, TGF-β is the only cytokine that is significantly elevated in patients' cerebrospinal fluid (CSF).[10] Besides TGF-β1, its functionally related isoform TGF-β2 is increased in the CSF as well.[11] However, the increase observed failed to reach significance.

In studies of mice that overexpressed TGF-β1 in the hippocampus, TGF-β1 was found to influence social interaction and depressive-related behavior.[12] It has not been assessed whether TGF-β modulates behavior independent of its role in Aβ production and clearance. Patients with AD show an impairment of memory that has been thought to result from alterations in the hippocampus, a critical brain area for memory processing.[13–15] Recent studies suggest that memory decline in AD could result from impairment of sleep-dependent memory consolidation.[16,17] In amnestic mild cognitive impairment, which proceeds AD, inadequate memory consolidation is associated with impaired sleep.[18] AD patients show disturbances of the circadian rhythm (for review, see Ref. 19). Daytime agitation, nighttime insomnia, and restlessness are among the common behavioral changes in AD.

It is hypothesized that the abnormal sleep–wake behavior in AD may be due to alterations of the molecular clock that maintains the circadian rhythm. The integration of the day and night variation is provided by the suprachiasmatic nucleus (SCN) of the hypothalamus, which receives light information from the retina and synchronizes the clock genes of other brain areas and peripheral organs. Every cell of the body contains functional feedback loops composed of positive transcriptional activators *Clock* and *Bmal1* and negative elements, including *Per* and cryptochrome (*Cry*). CLOCK and BMAL1 heterodimerize and bind to E-box enhancer sequences to promote transcription of the period genes *Per1* and *Per2*, and of the cryptochrome genes *Cry1* and *Cry2*, which themselves inhibit the activity of CLOCK-BMAL1. A second feedback loop, comprising REV-ERBα, controls the expression of the positive clock element *Bmal1*.[20,21]

In this study, we assessed the role of TGF-β2 on clock gene expression in fibroblasts and neuronal cells. We found that in CSF of AD patients, detectable concentrations of TGF-β2 profoundly inhibited the expression of *Period* genes, *rev-erbα*, and the clock-controlled genes *Dbp* and *Tef*.

Methods

Participants

Patients with mild cognitive impairment (MCI) and AD were recruited from the Memory Clinic at the Division of Psychiatry Research and Psychogeriatric Medicine, University of Zurich. Nondemented control subjects were recruited from the Departments of Neurology and Anesthesiology at the University Hospital Zurich. The study was approved by the local ethics committee. Written informed consent was obtained from all subjects before the investigation. Additional consent was obtained from caregivers in the case of dementia.

The cognitively impaired group (CI) was comprised of 7 (4 female) subjects with MCI (mean age 69.4 years ± 9.7, mean MMSE 26.7 ± 1.5) and 16 (8 female) subjects with AD (mean age 71.1 years ± 9.1, mean MMSE 17.8 ± 3.5). Patients with AD were diagnosed as "probable AD" according to the NINCDS-ADRDA criteria and "Alzheimer's dementia" according to ICD (F 00.0 and F 00.1) after extensive clinical work-up (medical history, clinical examination, ECG, MRI or CCT, blood and urine laboratory assessments).[22] The psychometric test battery followed the CERAD procedure, including verbal fluency, the Boston naming task, recall and delayed recall tests, and the Mini-Mental State Examination (MMSE). MCI subjects were diagnosed according to Petersen,[23] requiring documented impairment in one memory test and independent daily living. The group of nondemented controls (N = 10, 5 females, mean age 65.1 ± 7.3) consisted of subjects without evidence for cognitive disturbances by clinical examination and judgment of the physician in charge, who underwent lumbar puncture at the Department of Anesthesiology ($n = 7$) or Neurology ($n = 1$) or at the memory clinic ($n = 2$).

Cerebrospinal fluid sampling and quantification of TGF-β1 and TGF-β2 in CSF

Spinal taps were done between lumbovertebral body 3 and 4 or 4 and 5 using iodine solution (Braunol, B. Braun Medical AG, 6204 Sempach, Switzerland) as disinfectant and Sprotte-canula 21G × 31/2 (99 mm) for punctuation needles. The time point of punctuation was between 8 a.m. and 5 p.m. Six mL of cerebrospinal fluid were collected in 13 mL polypropylene tubes (round bottom, screw caps high density polypropylene) and immediately sampled in approximately 24 sterile caps (0.5 mL, Screw Cap Tubes, Art. Nr. SCT-050-C-S, Axygen Scientific, Union City, CA) with 250 µL volume each. TGF-β was measured using the Quantikine human TGF-β1 and TGF-β2 immunoassy from R&D (Abingdon, UK). To activate latent TGF-β1 to the immunoreactive form, acid activation and neutralization was performed according to the recommendations from R&D.

Clock gene expression in TGF-β2–treated cells

Murine fibroblast cells, NIH 3T3 (CRL-1658), were obtained from the American Type Culture Collection. To generate mouse embryonic fibroblasts (MEF), embryos of *Smad3* homozygous wild type (WT) or knockout (KO) from the same breeding were used at day 12 of gestation (E12). Liver and head were removed and the remaining tissue was digested with 0.25% trypsin-EDTA in Dulbecco's modified eagle medium (DMEM) for 30 min at room temperature. The dissociated cells were plated in DMEM containing 20% fetal calf serum (FCS). They were split 1:4 until they reached passage 10. Cells were used within passage 10–30. NIH 3T3 fibroblasts and MEFs were grown in DMEM (Gibco, Basel, Switzerland), supplemented with 10% phosphate-buffered saline (PBS) (PAA Laboratories, Pasching, Austria) and Glutamax (Gibco). For TGF-β2 treatment, cells were grown to confluency, and then the medium was replaced by serum-free DMEM/Glutamax with or without TGF-β2 (recombinant human TGF-β2; mammalian derived; 100–35B) from Peprotech (London, UK). After various time points of cell treatment, tissue culture plates were washed once with the ice-cold PBS solution and kept at −70 °C until the extraction of whole-cell RNA.

The mouse hippocampus neuronal cell line HT22 was obtained from David Schubert at the Salk Institute (La Jolla, CA). HT22 cells were plated in 12-well tissue cultures plates (100,000 cells per well) in DMEM with 10% FCS. Two days after plating, cultures were serum deprived for 1 hour. Thereafter, HT22 cells were treated with TGF-β2 for 4 hours.

Whole-cell RNA from cultured cells was extracted using TRIzol (Invitrogen; Life Technologies Europe, Zug, Switzerland) or peqGOLD RNAPure (peqLab) according to the manufacturer's instructions. Sub-

sequently, RNA was reverse-transcribed using random hexamers (Roche) and M-MuLV reverse transcriptase (Applied Biosystems). The cDNA equivalent to 20 ng of total RNA was PCR amplified in an ABI PRISM HT7900 detection system (PE-Applied Biosystems) using the TaqMan Universal PCR Master Mix (Applied Biosystems) and quantified as follows. Primers and probes for Taqman analysis were either purchased from Applied Biosystems or purchased from Microsynth (Balgach, Switzerland). With the exception of the *Runx3* and *Dec1* primers, all other primers have been described previously.[24] The primers used for *Runx3* were 5-ACCGCTTTGGAGACCTGCGCATG-3 and 5-CGCTGTAGGGGAAGGCGGCAGA-3, and for *Dec1*, 5-GAGACCCTGCGATCCTCCC-3 and 5-AGGTCTCCGTGCTCCAGCC-3. The relative levels of each RNA were calculated by the 2-$\Delta\Delta$Ct method (Ct standing for the cycle number at which the signal reaches the threshold of detection); *Gapdh* mRNA was used as a housekeeping gene. Each Ct value used for these calculations is the mean of two duplicates of the same reaction. Relative RNA levels are expressed as *x*-fold variations compared with untreated.

Statistical analysis

Statistical analyses were calculated using Student's *t*-test (GraphPad Prism version 4.0).

Results

TGF-β2 impairs the expression of clock genes

Previous reports indicate that *in vitro* cultures of fibroblasts provide an excellent tool to describe circadian expression of central clock and clock-controlled genes for at least three cycles.[25] For the study presented here, 3T3 fibroblasts were exposed to various doses of TGF-β2. This cytokine, rather than its isoform TGF-β1, was chosen because TGF-β2 is present in the CSF in higher concentrations than TGF-β1 (see below). However, both TGF-β1 and TGF-β2 bind to the same receptor TGF-βRII and share their biological activities. The expression of the clock-controlled gene *Dbp* was assessed after 24 hours by RT-PCR. TGF-β2 profoundly inhibited *Dbp* expression. A total of 75% inhibition of *Dbp* expression was seen with TGF-β2 100 pg/mL, and complete suppression of *Dbp* expression was achieved with TGF-β2 at 5 ng/mL (Fig. 1A). A statistically significant effect of TGF-β2 required

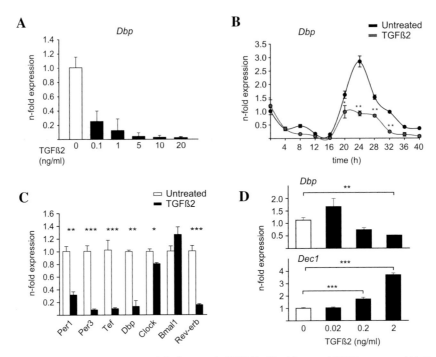

Figure 1. TGF-β2 suppresses the expression of clock genes in NIH3T3 fibroblasts and HT22 neurons. (A) Dose-dependent downregulation of *Dbp* gene expression. (B) TGF-β2 (1 ng/mL) reduces the amplitude of *Dbp* expression in synchronized NIH3T3 fibroblasts. (C) TGF-β2 (1 ng/mL)–induced inhibition of *Per1, Per3, Tef, Dbp, RevErbα*, and *Clock*, but not of *Bmal1*. (D) TGF-β2 inhibits the expression of *Dbp* in HT22 neuronal cells, the effect being paralleled by an upregulation of *Dec1*. All experiments were done in triplicate cultures. Data show the mean ± SEM. ***$P < 0.0005$; **$P < 0.005$; *$P < 0.05$.

its presence for 4 h in fibroblast cultures (data not shown). Next, we determined whether TGF-β2 alters the rhythmic expression of clock genes. Cultures were treated with TGF-β2 and thereafter RNA was extracted every 4 h. TGF-β2 did not alter the phasic expression of *Dbp*, but rather suppressed its maximum oscillation (Fig. 1B). To gain a more complete picture of the role of TGF-β2 in clock gene expression, the effect of the cytokine was also tested on *Per1, Per3, Tef, Rev-erbα*, and the central core clock genes *Clock* and *Bmal1*. Whereas TGF-β2 did not influence the expression of *Bmal1* and had only a minor effect on the expression of *Clock* (20% inhibition), the other clock genes analyzed were inhibited by >70% when compared with untreated cultures (Fig. 1C). TGF-β2 also inhibited the expression of *Dbp* in HT22 neuronal cells. This effect was dose dependent and paralleled by an increase of *Dec1* (Fig. 1D). When TGF-β2 was added in a concentration of 2 ng/mL, the suppression of *Dbp* was 49% and the induction of *Dec1* was 3.7-fold.

Smad3 signaling is only partially involved in the suppression of Per3 and Dbp by TGF-β2

Binding of TGF-β to TGF-βRII leads to phosphorylation of TGF-βRI. This step is followed by recruitment and phosphorylation of Smad2 and Smad3 that act together as heterodimers. In further experiments, we tested TGF-β2 on MEFs, which were established from *Smad3* gene knockout mice (Smad3$^{-/-}$). To verify SMAD3 deficiency at a functional level, the response of Smad3$^{-/-}$ MEFs to TGF-β2 was assessed by investigating the expression of *Runx3*, a well-established TGF-β target gene.[26] TGF-β2 failed to induce *Runx3* in Smad3$^{-/-}$ MEFs, but led to a profound increase in WT cells (Fig. 2A). Upon treatment of Smad3$^{-/-}$ MEFs with TGF-β2, the induced suppression of both *Dbp* and *Per3* expression was significantly less pronounced compared with WT MEFs (Fig. 2B and C). The percentage of TGF-β2–induced inhibition of expression of *Dbp* in wild-type and Smad3$^{-/-}$ MEFs was 85.8 ± 2.5% and 46.9 ± 8.1% ($P < 0.0001$), respectively. Likewise, the TGF-β2–induced inhibition of

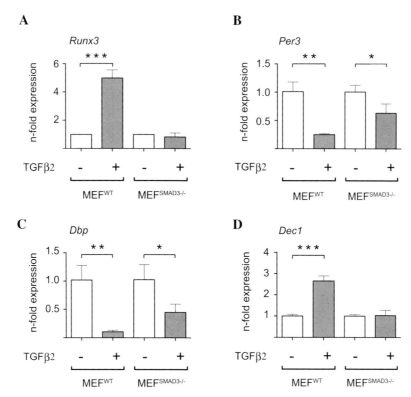

Figure 2. The suppression of expression of *Dbp* and *Per3* by TGF-β2 is not exclusively regulated by Smad3. (A and D) The upregulation of *Runx3* and *Dec1* by TGF-β2 is completely abolished in Smad3-deficient MEFs. (B and C) Downregulation of *Per3* and *Dbp* is only partly dependent on Smad3. All experiments were done in triplicate cultures. Data show the mean ± SEM. ***$P < 0.0005$; **$P < 0.005$; *$P < 0.05$.

expression of *Per3* was more pronounced in WT compared with Smad3$^{-/-}$ MEFs, the respective values being 63.8 ± 10.4% and 25.3 ± 7.3% ($P < 0.003$). However, the data show a significant Smad3-independent inhibitory effect of TGF-β2 on the expression of *Dbp* and *Per3* (Figs. 2B and C). The same results were seen when testing another eight MEF cell clones established from Smad3$^{-/-}$ mice (data not shown). A recent study showed TGF-β induces the basic helix-loop-helix protein *Dec 1,* an effect that is Smad3-dependent.[27,28] *Dec1* has been described to be involved in the expression of clock genes.[29] Our data confirm that TGF-β upregulates *Dec1* expression in WT MEFs, but not in Smad3$^{-/-}$ MEFs (Fig. 2D).

TGF-β1 and TGF-β2 concentrations in the CSF of AD patients

On the basis of the intriguing function of TGF-β2 to modulate the expression of clock genes, we elected to study CSF samples of patients with MCI ($n = 7$)

and AD ($n = 16$), the CSF of both patient groups hereby named CSF-CI (cognitively impaired (CI)). The reported concentration of TGF-β in CSF-CI varies significantly, ranging from between 0.2 ng/mL and 30 ng/mL.[30,31] The reasons for the variability are not entirely clear, but may be due to TGF-β being present in latent form. Activation of the latent form can be achieved by heat or acid treatment. To overcome variabilities due to CSF storage and freeze–thaw cycles, we acidified the CSF prior testing in a commercial ELISA system. As shown in Figure 3, TGF-β1 can be detected in the CSF-CI and in healthy controls (HC). The mean ± SD for TGF-β1 in CSF-CI was higher compared with HC, the concentrations being 51.09 ± 13.08 pg/mL and 36.16 ± 28.50 pg/mL, respectively. Since TGF-β1 and its isoform TGF-β2 are increased in the brain of AD patients, we assessed TGF-β2 in our ELISA system as well. In CSF-CI, the mean concentrations of TGF-β2 were higher compared with HC, the mean ± SD for CSF-CI and HC being 50.09 ± 28.43 and

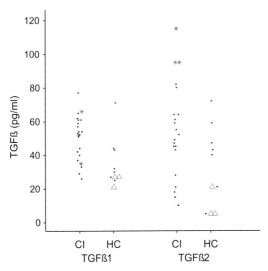

Figure 3. TGF-β1 and TGF-β2 in CSF of CI patients and HC. Stars indicate three CSF samples (two AD patients, one MCI patient) with highest concentrations of TGF-β2, and open triangles refer to the three CSF samples with the lowest TGF-β2 concentrations. For identification of these CSF samples, the same symbols were used in the data showing the concentrations of TGF-β1.

36.36 ± 27.31, respectively. The difference found did not reach statistical significance. When dissecting the CSF-CI into samples from AD and MCI patients, no differences in the concentration of TGF-β1 and TGF-β2 were found (data not shown). In two AD patients and one patient with MCI, the TGF-β2 concentrations were above 90 pg/mL, the mean ± 2 SD of HC. In these three CSF samples, the TGF-β2 CSF concentrations measured was between 95 pg/mL and 115 pg/mL.

CSF of CI patients neither suppresses the expression of clock genes nor induces the TGF-β target gene Dec1

Since in the three CSF samples outlined previously the TGF-β2 concentrations measured (around 0.1 ng/mL) were in the range of those required for significant suppression of the *Dbp* gene (Figs. 1B and Fig. 3), we evaluated the impact of CSF-CI on the expression of clock genes. To overcome the problem of small volumes of the CSF samples used, we formed a pool with the three CSF samples with the highest TGF-β2 concentrations (see symbol * in Fig. 3), This pool failed to suppress the expression of *Dbp*, *Per1*, and *Per3* (data not shown). Since the CSF pools had to be diluted fourfold in

the assay, the absence of effects of the CSF pool may be due to TGF-β2 concentrations that are under the detection limit. Alternatively, CSF samples may harbor factors that hinder TGF-β2 to act on the expression of clock genes. When adding TGF-β2 to CSF samples derived from CI patients with low TGF-β2, the CSF samples supplemented with the higher TGF-β2 concentration (2 ng/mL) were found to inhibit the expression of *Dbp* and to enhance the expression of *Dec1* (Fig. 4). The effect on *Dbp* expression did not become visible when the CSF samples were supplemented with less TGF-β2 (0.1 ng/mL). These data indicate that the absence of downregulation of clock genes by the CSF from patients with CI is not due to the presence of factors in the CSF, which would interfere with the biological function of TGF-β2 to modulate clock gene expression.

Discussion

In AD, sleep efficiency is reduced and sleep–wake rhythms disturbed (for review, see Ref. 19). Disruption of sleep–wake rhythm is the most frequent reason for nursing home placement.[32] Interestingly, the rhythmic expression of the clock genes *Bmal1*, *Cry1*, and *Per1* was diminished in preclinical and clinical AD.[33] In our study, we focused on the modulation of clock gene expression by TGF-β, because according to a recent meta-analysis on cytokines in AD, TGF-β1 is the only cytokine identified to be increased in the CSF of AD patients; TGF-β1 and TGF-β2 are expressed on astrocytes and neurons in AD patients, and inhibition of the TGF-β–signaling pathways in mice, which overexpress Aβ, mitigates the disease (references are given in the introduction). Using fibroblasts and HT22 neuronal cells, we found that TGF-β2 suppressed the expression of *Per1*, *Per2*, and *Rev-erbα* and the clock-controlled genes *Dbp* and *Tef*; however, the expression of the *Bmal1* gene was not significantly altered. An analogous effect has been described in studies on TNF-α, which impairs binding of Clock-Bmal1 heterodimers to the E-box sequences of *Dbp* and *Rev-erbα*.[34]

The mechanisms involved in the TGF-β–induced inhibition of E-box–regulated clock genes are not yet clear. In Ras-1 fibroblasts, which stably express a luciferase reporter under the regulation of a 0.3-kb *Bmal1* promoter, TGF-β1 leads to a phase-shift of the circadian oscillations of the

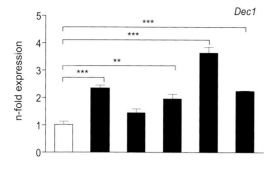

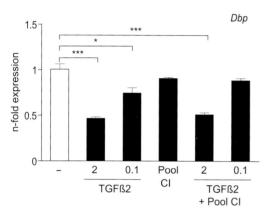

Figure 4. CSF from patients with CI does not impair the effect of TGF-β2 on the expression of *Dbp* and *Dec1*. HT22 neuronal cells were treated either with TGF-β2 (2 ng/mL or 0.1 ng/mL) or with a CSF pool that was derived from three CI patients with the lowest TGF-β2 concentrations (pool CI). Before testing, the pool, CI was supplemented with TGF-β2 (2 ng/ul or 0.1 ng/mL). The experiments were done in triplicate cultures. Data show the mean ± SEM. ***$P < 0.0005$; **$P < 0.005$; *$P < 0.05$.

bioluminescence signal.[27] This effect of TGF-β1, which is mediated by activation of the activin receptor-like kinase, was also associated with suppression of *Dbp* and *Rev-erbα* as well as an increased expression of the *Dec1*. *Dec1* (also known as Stra13), which is induced by TGF-β, is a basic helix-loop-helix protein that can act as a repressor of gene transcription.[28] Our data show that TGF-β2 not only inhibits *Dbp* and *Rev-erbα* in fibroblasts, but also leads to a prominent decrease of *Per1*, *Per3*, and *Tef* at the time point of an unaltered expression of *Bmal1*. Furthermore, our timecourse analysis of *Dbp* expression revealed that TGF-β2 inhibits the amplitude of *Dbp* oscillation without causing a phase-shift of its expression. TGF-β-induced signaling involves the activation of Smad proteins. As reported by others, we find the induction of *Dec1* to be dependent on ex-

pression of Smad3 in TGF-β2 treated fibroblasts.[27] However, despite the absence of *Dec1* expression in Smad3$^{-/-}$ MEFs, TGF-β2 was still able to suppress the expression of *Per3* and *Dbp*—the effect, however, being less pronounced. This is of note since TGF-β1 causes a phase-shift of the expression of *Dbp* and of *Per1 in vivo* in mice in a *Dec1*-dependent manner.[27] Smad-independent effects of TGF-β1 occur through various pathways, including the Ras-extracellular signal-regulated kinase (Erk), TGF-β1-activated kinase-mitogen-activated protein kinase (MAPK), and phosphatidylinositol 3-kinase (PI3K)-Akt pathways.[35]

In a meta-analysis of cytokines detected in AD, CSF TGF-β1 was found to be the only cytokine elevated.[10] This contrasts with levels of TNF-α, IL-1, and IL-6, which were not increased. The analysis comprised five studies with a total of 113 AD patients and 114 age-matched controls. All five studies investigated the CSF concentrations of TGF-β1, but not of TGF-β2. In addition, the amount of TGF-β1 in CSF correlated with the MMSE scores.[30] TGF-β2 is synthesized in AD by hippocampal and cortical neuronal cells and upregulated in familial AD cases, with presenilin mutations in astrocytes around senile plaques and in neurons with neurofibrillary tangles.[7] In homogenates of brain tissue of AD, TGF-β2 is 3.2 times the average level of control samples.[7] TGF-β2 has only been assessed in the CSF in one study.[11] There was a trend for TGF-β2 to be elevated in AD, which failed to achieve significance. TGF-β2 was increased compared with controls, but the effect was not statistically significant. Overexpression of TGF-β in neurons and plaques of AD patients may, on a quantitative level, not be mirrored in corresponding changes in the CSF TGF-β content. The comparison between CNS tissue and CSF is complex because binding of TGF-β to TGF-βRII in brain tissue may lead to an underestimation of TGF-β in the CSF. Our data show that 2 out of 16 AD patients and 1 out of 7 patients with MCI had TGF-β2 CSF concentrations above 90 pg/mL. This is remarkable because a TGF-β2 concentration of 100 pg/mL led to a 75% inhibition of *Dbp* expression. However, a pool of the three CSF samples with the highest TGF-β2 levels did not mimic the effects on clock genes as seen with recombinant TGF-β2. When supplementing CSF samples with TGF-β2, the effect of the cytokine to inhibit the expression of *Dbp* and to enhance the expression of *Dec1* in

HT22 neuronal cells becomes detectable. These experiments indicate that there are no factors in the CSF that would override the activity of TGF-β2 on clock gene expression, and the dilutions of the CSF samples required in the clock gene assay are too high to allow for the functional detection of TGF-β2 in the CSF.

Here, we document the efficacy of TGF-β2 to downregulate central clock genes (*Period* genes) and clock-controlled genes (*Dbp, Tef*), but not the master clock gene *Bmal1*. The significance of these findings is provided by the observation that patients with AD have disturbances of the circadian rhythm,[19] and the concept that the timing of sleep and wakefulness and sleep structures results from the interaction of a circadian and a sleep–wake-dependent homeostatic process.[36] Findings in gene KO mice with inactivation of individual clock genes led to the hypothesis that circadian clock genes, in addition to controlling circadian timing of sleep, may also be important to sleep-regulatory processes.[37] For example, while total sleep time remains constant in *Per1* and *Per2* mutant mice, the distribution of sleep is affected by the mutation (see Ref. 37). Collectively, the data presented indicate that TGF-β–induced dysregulation of clock gene expression may play a role in abnormal sleep–wake rhythms in Alzheimer's disease.

Acknowledgments

This research was supported by grants from the Swiss National Science Foundation (Project 310030_141055/1 to H.M. and A.F.) and the Lotex Foundation (to A.F.).

Conflicts of interest

The authors declare no conflicts of interest.

References

1. Lesne, S., F. Docagne, C. Gabriel, *et al.* 2003. Transforming growth factor-beta 1 potentiates amyloid-beta generation in astrocytes and in transgenic mice. *J. Biol. Chem.* **278:** 18408–18418.

2. Wyss-Coray, T., C. Lin, F. Yan, *et al.* 2001. TGF-beta1 promotes microglial amyloid-beta clearance and reduces plaque burden in transgenic mice. *Nat. Med.* **7:** 612–628.

3. Wyss-Coray, T., E. Masliah, M. Mallory, *et al.* 1997. Amyloidogenic role of cytokine TGF-beta1 in transgenic mice and in Alzheimer's disease. *Nature* **389:** 603–606.

4. Town, T., Y. Laouar, C. Pittenger, *et al.* 2008. Blocking TGF-beta-Smad2/3 innate immune signaling mitigates Alzheimer-like pathology. *Nat. Med.* **14:** 681–687.

5. Hashimoto, Y., T. Chiba, M. Yamada, *et al.* 2005. Transforming growth factor beta2 is a neuronal death-inducing ligand for amyloid-beta precursor protein. *Mol. Cell Biol.* **25:** 9304–9317.

6. Hashimoto, Y., M. Nawa, T. Chiba, *et al.* 2006. Transforming growth factor beta2 autocrinally mediates neuronal cell death induced by amyloid-beta. *J. Neurosci. Res.* **83:** 1039–1047.

7. Noguchi, A., M. Nawa, S. Aiso, *et al.* 2010. Transforming growth factor beta2 level is elevated in neurons of Alzheimer's disease brains. *Int. J. Neurosci.* **120:** 168–175.

8. Lee, H.G., M. Ueda, X. Zhu, *et al.* 2006. Ectopic expression of phospho-Smad2 in Alzheimer's disease: uncoupling of the transforming growth factor-beta pathway? *J. Neurosci. Res.* **84:** 1856–1861.

9. Beynon, A.L., A.N. Coogan & J. Thome. 2009. Age and time of day influences on the expression of transforming growth factor-beta and phosphorylated SMAD3 in the mouse suprachiasmatic and paraventricular nuclei. *Neuroimmunomodulation* **16:** 392–399.

10. Swardfager, W., K. Lanctot, L. Rothenburg, *et al.* 2010. A meta-analysis of cytokines in Alzheimer's disease. *Biol. Psychiatr.* **68:** 930–941.

11. Vawter, M.P., O. Dillon-Carter, W.W. Tourtellotte, *et al.* 1996. TGFbeta1 and TGFbeta2 concentrations are elevated in Parkinson's disease in ventricular cerebrospinal fluid. *Exp. Neurol.* **142:** 313–322.

12. Depino, A.M., L. Lucchina & F. Pitossi. 2011. Early and adult hippocampal TGF-β1 overexpression have opposite effects on behavior. *Brain Behav. Immun.* **25:**1582–1591.

13. Morris, J.C. 1996. Classification of dementia and Alzheimer's disease. *Acta. Neurol. Scand. Suppl.* **165:** 41–50.

14. Nadel, L. & M. Moscovitch. 1997. Memory consolidation, retrograde amnesia and the hippocampal complex. *Curr. Opin. Neurobiol.* **7:** 217–227.

15. Squire, L.R. & P. Alvarez. 1995. Retrograde amnesia and memory consolidation: a neurobiological perspective. *Curr. Opin. Neurobiol.* **5:** 169–177.

16. Rauchs, G., B. Desgranges, J. Foret & F. Eustache. 2005. The relationships between memory systems and sleep stages. *J. Sleep Res.* **14:** 123–140.

17. Rauchs, G., M. Schabus, S. Parapatics, *et al.* 2008. Is there a link between sleep changes and memory in Alzheimer's disease? *Neuroreport* **19:** 1159–1162.

18. Westerberg, C.E., E.M. Lundgren, S.M. Florczak, *et al.* 2010. Sleep influences the severity of memory disruption in amnestic mild cognitive impairment: results from sleep self-assessment and continuous activity monitoring. *Alzheimer Dis. Assoc. Disord.* **24:** 325–333.

19. Weldemichael, D.A. & G.T. Grossberg. 2010. Circadian rhythm disturbances in patients with Alzheimer's disease: a review. *Int. J. Alzheimers Dis.* 1–9.

20. Hastings, M.H. 2003. Circadian clocks: self-assembling oscillators? *Curr. Biol.* **13:** R681–R682.

21. Schibler, U. 2009. The 2008 Pittendrigh/Aschoff lecture: peripheral phase coordination in the mammalian circadian timing system. *J. Biol. Rhythms* **24:** 3–15.

22. McKhann, G., D. Drachman, M. Folstein, *et al.* 1984. Clinical diagnosis of Alzheimer's disease: report of the NINCDS-ADRDA Work Group under the auspices of Department

of Health and Human Services Task Force on Alzheimer's Disease. *Neurology* **34:** 939–944.

23. Petersen, R.C., G.E. Smith, S.C. Waring, *et al.* 1999. Mild cognitive impairment: clinical characterization and outcome. *Arch. Neurol.* **56:** 303–308.

24. Petrzilka, S., C. Taraborrelli, G. Cavadini, *et al.* 2009. Clock gene modulation by TNF-alpha depends on calcium and p38 MAP kinase signaling. *J. Biol. Rhythms* **24:** 283–294.

25. Nagoshi, E., C. Saini, C. Bauer, *et al.* 2004. Circadian gene expression in individual fibroblasts: cell-autonomous and self-sustained oscillators pass time to daughter cells. *Cell* **119:** 693–705.

26. Ito, Y. & K. Miyazono. 2003. RUNX transcription factors as key targets of TGF-beta superfamily signaling. *Curr. Opin. Genet Dev.* **13:** 43–47.

27. Kon, N., T. Hirota, T. Kawamoto, *et al.* 2008. Activation of TGF-beta/activin signalling resets the circadian clock through rapid induction of *Dec1* transcripts. *Nat. Cell Biol.* **10:** 1463–1469.

28. Zawel, L., J. Yu, C.J. Torrance, *et al.* 2002. *Dec1* is a downstream target of TGF-beta with sequence-specific transcriptional repressor activities. *Proc. Natl. Acad. Sci. U.S.A.* **99:** 2848–2853.

29. Honma, S., T. Kawamoto, Y. Takagi, *et al.* 2002. *Dec1* and *Dec2* are regulators of the mammalian molecular clock. *Nature* **419:** 841–844.

30. Rota, E., G. Bellone, P. Rocca, *et al.* 2006. Increased intrathecal TGF-beta1, but not IL-12, IFN-gamma and IL-10 levels in Alzheimer's disease patients. *Neurol. Sci.* **27:** 33–39.

31. Tarkowski, E. 2002. Cytokines in dementias. *Curr. Drug Targets Inflamm. Allergy* **1:** 193–200.

32. Lebert, F., F. Pasquier & H. Petit. 1996. Sundowning syndrome in demented patients without neuroleptic therapy. *Arch. Gerontol. Geriatr.* **22:** 49–54.

33. Wu, Y.H., D.F. Fischer, A. Kalsbeek, *et al.* 2006. Pineal clock gene oscillation is disturbed in Alzheimer's disease, due to functional disconnection from the "master clock." *FASEB J.* **20:** 1874–1876.

34. Taraborrelli, C., S. Palchykova, I. Tobler, *et al.* 2011. TNFR1 is essential for CD40, but not for lipopolysaccharide-induced sickness behavior and clock gene dysregulation. *Brain Behav. Immun.* **25:** 434–442.

35. Eger, A., K. Aigner, S. Sonderegger, *et al.* 2005. DeltaEF1 is a transcriptional repressor of E-cadherin and regulates epithelial plasticity in breast cancer cells. *Oncogene* **24:** 2375–2385.

36. Borbely, A. 1982. Endogenous sleep-promoting substances. *Trends Pharmacol. Sci.* **3:** 350-.

37. Franken, P., D. Chollet & M. Tafti. 2001. The homeostatic regulation of sleep need is under genetic control. *J. Neurosci.* **21:** 2610–2621.

Ann. N.Y. Acad. Sci. ISSN 0077-8923

ANNALS OF THE NEW YORK ACADEMY OF SCIENCES

Issue: *Neuroimmunomodulation in Health and Disease*

Role of interleukin-6 in stress, sleep, and fatigue

Nicolas Rohleder,[1] Martin Aringer,[2] and Matthias Boentert[3]

[1]Department of Psychology and Volen National Center for Complex Systems, Brandeis University, Waltham, Massachusetts. [2]Medical Clinic, University Carl Gustav Carus, Dresden, Germany. [3]Department of Neurology, University Hospital Muenster, Muenster, Germany

Address for correspondence: Nicolas Rohleder, Assistant Professor, Department of Psychology & Volen National Center for Complex Systems, Brandeis University, 415 South Street, MS062 P.O. Box 549110, Waltham, MA, 02454. rohleder@brandeis.edu

Chronic low-grade inflammation, in particular increased concentrations of proinflammatory cytokines such as interleukin (IL)-6 in the circulation, is observed with increasing age, but it is also as a consequence of various medical and psychological conditions, as well as life-style choices. Since molecules such as IL-6 have pleiotropic effects, consequences are wide ranging. This short review summarizes the evidence showing how IL-6 elevations in the context of inflammatory disease affect the organism, with a focus on sleep-related symptoms and fatigue; and conversely, how alterations in sleep duration and quality stimulate increased concentrations of IL-6 in the circulation. Research showing that acute as well as chronic psychological stress also increase concentrations of IL-6 supports the notion of a close link between an organism's response to physiological and psychological perturbations. The findings summarized here further underscore the particular importance of IL-6 as a messenger molecule that connects peripheral regulatory processes with the CNS.

Keywords: interleukin-6; rheumatoid arthritis; stress; sleep; fatigue

Introduction

Cytokines primarily are messenger molecules of the immune system and fulfill important functions in the communication between different immunologically active cells and tissues. As our understanding of these communication pathways deepens, it becomes increasingly clear that cytokines also transmit information to nonimmunological tissues. Cytokines are secreted not only by immune cells in the strict sense but also by endothelial, epithelial, and even glial cells.[1] Cytokines such as interleukin-1 (IL-1), tumor necrosis factor α (TNF-α), and IL-6 are known to exert both paracrine and endocrine effects.[2] Thus, they mediate not only local but systemic responses to physiological or pathological stimuli. The acute phase reaction (APR) exemplifies cytokine effects on various nonimmunological tissues: proinflammatory cytokines target the CNS (the hypothalamus in particular) and trigger fever, reduced activity, and sleep. Although locomotion is reduced, protein breakdown within skeletal musculature is markedly increased. In the liver, APR proteins are synthesized

and iron is redistributed to the intracellular compartment. In summary, these processes create an environment that is unfavorable for microbial agents and enables the host to better cope with microbial invasion. However, far beyond APR induction, cytokines participate in the regulation of extremely complex biological processes, including inflammation, physical, and psychological stress, sleep, and recovery. The CNS, for example, has been shown to receive information from the immune system via cytokines, and the CNS likewise affects peripheral cytokine concentrations, such as through activation of the stress system (e.g., Refs. 3 and 4).

Interleukin-6 is classified as one of the proinflammatory cytokines that stimulate B cell function.[5] Its receptor consists of a specific α-chain, CD126,[6] and two chains of gp130,[7] which are share with the receptor for IL-11, leukemia inhibitory factor-1 (LIF-1), ciliary neutrophic factor, oncostatin M, and cardiotrophin-1.[8] The main signal transduction molecule of the IL-6 receptor is signal transducer and activator of transcription-3 (STAT3),[9] which is bound by gp130 and

doi: 10.1111/j.1749-6632.2012.06634.x

phosphorylated by Janus kinase 2 (Jak2).[10] Upon tyrosine phosphorylation, STAT3 homodimerizes, shuttles to the nucleus, and subsequently functions as a transcription factor. While leukocytes and hepatocytes carry the IL-6 receptor α-chain, many other cells do not but still respond to IL-6, particularly under inflammatory conditions.[11] This is possible via binding of soluble (mostly shedded) IL-6Rα to membrane-bound gp130, a process called *trans*-signaling.[12]

While many cytokines have been shown to play a role outside the immune system, IL-6 is the cytokine with the most documented effects on nonimmunological tissues. IL-6 has therefore been termed a pleiotropic or "endocrine" cytokine[13] because it is secreted by a variety of nonimmune cells, for example, adipocytes[14] and endothelial cells.[15] IL-6 has also been nicknamed the "cytokine for gerontologists,"[16] because of its role in many age-related diseases.[17] IL-6, together with C reactive protein (CRP), is one of the most important biomarkers of chronic or systemic low-grade inflammation, and there is now strong evidence of the predictive power of plasma IL-6 and CRP concentrations for later life morbidity and mortality (e.g., Ref. 18). IL-6 is strongly induced by IL-1β. Circulating IL-6, together with TNF-α and IL-1, is required for APR induction.[19] In particular, IL-6 supports IL-1 in the generation of fever,[20] and it enhances hepatic production of CRP and serum amyloid P component.[21] At the same time, IL-6 also exhibits anti-inflammatory effects by controlling levels of TNF-α, IFN-γ, and GM-CSF.[22] IL-6–induced production of CRP (which is strongly enhanced by IL-1) mediates microbial opsonization, complement activation, and phagocytosis.[23] Thus, IL-6 is not only involved in ostensibly detrimental aspects of APR, such as fever, fatigue, and EDS, it also exerts protective effects on host defense.

Numerous studies have shown that IL-6 is crucial for interaction between the immune system and the CNS in inflammatory disease, particularly for mediating common clinical symptoms like fatigue, sleep disturbances, and excessive daytime sleepiness. IL-6 is involved in normal sleep regulation and plays a role in both acute and chronic stress responses and also in the physiological processes underlying recovery.

The aim of this short review is to summarize the role of IL-6 in the related conditions of inflammatory disease, sleep, fatigue, and stress. In its first section, the paper will focus on rheumatoid arthritis as a clinical representative of inflammatory disease. In the following sections, we will summarize the role of IL-6 in fatigue and sleep, as well as in stress reaction. Since this review strongly focuses on IL-6, the specific effects of TNF-α and IL-1β will not be discussed in detail.

IL-6 and IL-6 blockade in rheumatoid arthritis

Rheumatoid arthritis (RA) is an inflammatory system disease. While erosive polyarthritis is the characteristic clinical sign of RA,[24] the problem is not limited to swelling, pain, and consecutive destruction of joints. Indeed, RA is also associated with significant mortality, which is highly correlated with the level of inflammation,[25] leading to increased prevalence of atherosclerosis and lymphoma, and may also be associated with RA vasculitis, or even amyloidosis.

The acute phase reaction has long been known to be of importance in estimating RA activity. While the 28-joint disease activity score (DAS28) used erythrocyte sedimentation rate (ESR) as a marker of inflammation, CRP is currently considered the best clinical marker.[26] CRP is directly induced by IL-6,[27] which was therefore long thought to play a relevant role in RA. In fact, tocilizumab, a therapeutic humanized monoclonal antibody against the IL-6 receptor that actively blocks IL-6 from binding, is FDA- and EMA-approved for the treatment of RA refractory to standard therapy with disease modifying antirheumatic drugs (DMARDs).[28–30] The biological response modifier tocilizumab not only effectively blocks joint inflammation, it also reduces radiographic joint damage.[31] This approach of directly targeting IL-6 allows for the examination of IL-6 effects beyond the direct effects on arthritis.

Data show that under the tocilizumab dose of 8 mg/kg body weight licensed in Europe, CRP levels drop rapidly and then constantly stay low at (near) normal levels in most patients.[30] At the same time, patients started to report a reduction of fatigue,[30] which many RA patients experience as incapacitating. In fact, in the TAMARA open-label trial, patients documented their fatigue on a daily basis.[32] Within days, patients experienced meaningful reductions in fatigue, reaching approximately half of the total benefit within a week. However, further analysis revealed that the maximum benefit was associated with remission,[33] suggesting that IL-6 is not the only molecule conferring RA-associated fatigue.

One aspect most probably involved in RA-associated fatigue is anemia associated with chronic disease, which is manifest in approximately one third of patients[34] and well-associated with IL-6.[35] Anemia is not only well known to go hand in hand with fatigue, but also with RA physical disability.[34] Indeed, low hemoglobin levels are found to rapidly increase under IL-6 receptor blockade with tocilizumab.[36] A mechanical explanation for this observation is that IL-6 drives hepcidin, which blocks ferroportin-mediated cellular iron transfer.[37,38]

Moreover, at approximately the same rate of improvement as with the patient estimates of fatigue and of RA disease activity were patient estimates of pain decreased.[32] Pain, and pain at night in particular, will often lead to impaired sleep quality, as will the morning stiffness typical of RA. RA-associated pain is mediated by other cytokines such as TNF-α, and inhibition of TNF-α receptor leads to rapid alleviation of pain in RA patients even before anti-inflammatory effects on the joints can be detected.[39] Pain-related sleep disturbances likely add to fatigue in active RA, and early alleviation of fatigue in RA patients treated with tocilizumab or etanercept may in part be attributed to pain reduction. However, fatigue in RA is more complex.[40] As shown in the following section, IL-6 also exerts direct effects on both sleep and fatiguability, and is intimately involved in the stress response.

IL-6, sleep, and fatigue

Normal sleep has been conceptualized as the interaction of the homeostatic and the circadian processes as postulated by Borbély. The homeostatic process ("S") describes the build-up of sleep pressure dependent on the time spent awake, with sleep leading to a rapid and substantial reduction of sleep drive.[41]

The circadian process ("C") comprises the alternating rhythmicity of wakefulness and sleep within a period of little more than 24 hours, which we are all familiar with. Its pacemaker has been located in the suprachiasmatic nucleus of the hypothalamus, which is regulated by the light–dark cycle, and which itself synchronizes various physiological rhythms, such as body temperature

and secretion of growth hormone, cortisol, and melatonin.[42]

Cytokines play a role in sleep regulation under physiological and pathological conditions. It is a long-standing clinical observation that acute or chronic infections, inflammatory diseases, and malignancies, may all induce sleep-related symptoms, such as reduced sleep quality, fatigue, and excessive daytime sleepiness (EDS).

Although often used in a similar sense by both patients and physicians, the terms EDS and fatigue can clearly be distinguished:[43] EDS is characterized by an abnormal sleep propensity during daytime, whereas fatigue describes a persistent feeling of physical or mental exhaustion, which is not accompanied by sleepiness, and which sleep cannot alleviate. Both clinical entities may be present in the same patient, and both have been shown to occur in association with the release of proinflammatory cytokines.

There is increasing evidence that chronic sleep loss, insomnia, sleep disturbances, and even abnormal duration of the main sleep period are correlated with adverse health outcomes and increased mortality.[44–47] Associated conditions include cardiovascular disease, diabetes, and obesity. In addition, sufficient sleep is the prerequisite of immunocompetence, and disordered sleep causes susceptibility to infection.[48,49]

It may be a subtle but persistent inflammatory response that builds the link between sleep impairment and increased morbidity. Both acute and chronic reduction of sleep duration and quality have been shown to induce oversecretion of proinflammatory cytokines.[50–52] IL-6 appears to be one of the most important cytokines that mediate the rapid interplay between the immune system and CNS function. It has been proposed to be a "sleep factor" that enhances sleep drive in accordance to the circadian rhythm.[53]

Serum levels of IL-6 reflect circadian and even ultradian cycles, as they are generally lower during daytime and higher during the night, with two peaks occurring at about 6:00–8:00 p.m. and 4:00–6:00 a.m.[54] During sleep, serum levels are high in sleep stages N1, N2, and REM, but lower in N3 or slow wave sleep (SWS). It is interesting that increases of IL-6 serum levels are highest when sleep occurs, while delay of sleep onset postpones the IL-6 secretion profile.[55]

Moreover, onset of sleep strongly increases serum levels of the soluble IL-6 receptor (sIL-6R), which enables IL-6 effects on nonimmune cells, both in the CNS and in peripheral organs.[56] Thus, IL-6 signaling is not only regulated in accordance with the sleep–wake cycle, but is also directly influenced by sleep itself. Preliminary evidence suggests that IL-6, in turn, may modify the circadian rhythm on a molecular level by activating period gene 1 (Per 1) transcription.[57] In contrast, Cavadini *et al.* found that expression of clock genes is reduced by IL-1β and TNF-α, but not by IL-6.[58] Thus, the specific effects of IL-6 on sleep regulation are still under debate, but alteration of clock gene functions by proinflammatory cytokines may be the molecular basis of fatigue and EDS in inflammatory disease.

Sleep deprivation leads to significant elevation and peaking of IL-6 serum levels the following day.[59] Serum levels can be reduced again by napping during daytime.[60] Total sleep loss over several days has been shown to persistently increase daytime serum levels of IL-6 and the sIL-6R.[61] SWS rebound in the first postdeprivation night is associated with undersecretion of IL-6.[53] In summary, IL-6 appears to mediate both sleep drive and recovery after sleep deprivation.

Exogenous administration of IL-6 has been shown to exert somnogenic effects in one animal study.[62] Other proinflammatory cytokines, such as IL-1β and TNF-α, clearly enhance SWS and prolong sleep in rats.[63–65] IL-1β effects on SWS have been shown to be independent of IL-6 in IL-6–deficient transgenic mice.[20] In healthy individuals, i.v. administration of IL-6 significantly alters sleep structure and promotes fatigue and several aspects of sickness behavior.[66] In contrast to normal sleep architecture, SWS is decreased in the first half of the night and increased in the second half after nighttime IL-6 administration. REM sleep is reduced throughout the night. In addition, exogenous IL-6 activates the hypothalamic–pituitary–adrenal (HPA) axis leading to transient, nonphysiological hypercortisolemia during the first hours of sleep, which may account for the reduction of sleep quality during this period.[67]

Moreover, IL-6 serum levels are elevated in patients with disorders of EDS, such as narcolepsy, obstructive sleep apnea syndrome (OSAS), and idiopathic hypersomnia.[52,68,69] In narcoleptic patients, TNF-α and growth hormone levels are also increased, indicating that dysregulation of immune and endocrine responses may contribute to EDS and reduced sleep quality.[68] In OSAS, intermittent hypoxemia mainly activates proinflammatory cytokines, including IL-6, which contributes to EDS in a synergistic manner with desaturation and sleep disruption. IL-6, TNF-α, and IL-1 are potential mediators of cardiovascular morbidity in OSAS.[52,70–72] Serum levels of sIL-6R have been shown to reflect disease severity in OSAS.[73]

The effects of IL-6 inhibition by tocilizumab on sleep and fatigue have been investigated by only a few studies so far. In the TAMARA study population, tocilizumab significantly reduced self-reported fatigue within 4 weeks of treatment.[32] The effect proved stable over 24 weeks. In the OPTION study, inhibition of IL-6 action in RA patients also led to significant reduction of self-reported fatigue compared to controls, but dose-dependent improvement was only 3–6% on the FACIT-fatigue score.[25] Sleep quality and EDS were not subject to both studies. Fragiadaki *et al.* reported robust improvement of self-reported sleepiness, sleep quality, and fatigue in 13 female RA patients receiving tocilizumab in a standard dosage.[74] Therapeutic effects of IL-6 blockade on sleep and daytime function showed no correlation to disease activity, suggesting that sleep-related symptoms and fatigue in RA may be mediated by direct cytokine effects on the CNS rather than ostensible factors such as joint disease and pain. As far as fatigue is concerned, the clinical relevance of this observation is still unclear because effect size appears to be small in studies testing both tocilizumab and other biotherapeutic agents in RA patients.[75] In addition, fatigue reduction under IL-6 blockade is probably highest during the first months of treatment, with stagnation or even a slight increase of self-reported fatigue at 24 weeks.[27]

Interestingly, alleviation of fatigue appears to be one of the first beneficial effects RA patients may experience under treatment with biotherapeutic agents blocking cytokine action. Studies on both etanercept and tocilizumab showed reduction of fatigue within days and weeks, which is well before anti-inflammatory effects on joint disease alone could account for this effect.[25,27,39,67] This finding suggests direct CNS effects of cytokine receptor blockade with regard to fatigue as it has already been shown for pain perception in RA patients.[39] In addition, it becomes even more clear that, among other factors, different cytokines act synergistically in the development and maintenance of fatigue in inflammatory disease.

IL-6 under conditions of stress

The effect of acute stress on IL-6 has been studied in several different ways. An important distinction to be made is between stress-induced changes of IL-6 in plasma or serum, and changes in mitogen-induced *in vitro* production of IL-6 in whole blood or cell cultures. We will only discuss the former here.

One line of evidence showing acute responses of plasma IL-6 comes from the literature on physiological effects of exercise. Acute physiological stress in the form of exercise of different types and duration is associated with marked increases of IL-6 in plasma, for example, after a 25-min treadmill exercise at 90% VO_2 max.[76] IL-6 responses to exercise seem to be mainly driven by release of IL-6 from muscles (for a summary see Ref. 77). However, there is evidence that this exercise-induced release is under stress system control, since IL-6 increase was found to correlate with catecholamine increases, and to be suppressed by glucocorticoid application.[76]

Related to these findings, there is now compelling evidence that acute stress of purely psychological nature can induce profound increases in circulating IL-6. This has been demonstrated in laboratory animals after open field exposure,[78] foot shock, restraint, and exposure to a conditioned aversive stimulus.[79] Similarly, studies in humans have shown that a variety of laboratory stress paradigms have the potential to induce increases in circulating IL-6, as summarized in a set of meta-analyses by Steptoe *et al.*[80] Including 18 studies published between 1993 and 2006, they found robust evidence of stress-induced increases of plasma IL-6, but not of TNF-α, and only a trend for CRP. Increases of IL-6 have been found to be higher if samples are taken after a longer lag period post stress, resulting from slow IL-6 stress responses.

Later studies have helped to better understand complex determinants of the stress-induced IL-6 increase: higher IL-6 increase has been documented in individuals with lower income,[81] lower physical fitness,[82] in participants of a meditation study who self-reported lower meditation practice,[83] and was found positively correlated with anger and anxiety related to the laboratory stress task.[84] In contrast, IL-6 responses to acute stress were not related with self-esteem[85] or yoga practice.[86] One recent study

did not find an IL-6 response 2 hours after a laboratory stress task.[87] Stress-induced increases in peripheral inflammation further seem to be elevated in in individuals exposed to early life adversity[88] or in combination of early life adversity and depression.[89]

The mechanism underlying the increase of circulating IL-6 after acute stress is still not well understood. IL-6 is secreted by nonimmunological and immune tissues, and, as documented in the case of exercise, these nonimmune tissues might explain much of the additional IL-6 in the circulation.[77] DNA-binding activity of the inflammatory transcription factor NF-κB has been found to increase in the same situations that lead to IL-6 increase, but more rapidly and with faster recovery than IL-6.[89,90] Such NF-κB stress responses appear to be under the control of sympathetic and HPA stress hormones,[90,91] similar to IL-6 increases.[76] Therefore, IL-6 could be driven by stimuli inducing NF-κB. However, since the NF-κB signaling pathway is not limited to immune cells, understanding this mechanism does not reveal any information about tissue sources. Several studies in both humans and animals have shown that i.v. administration of norepinephrine without stress increases circulating IL-6,[92–94] but that beta-blockade was unsuccessful in suppressing the IL-6 increase in human[95] and animal studies.[79] In the latter study, adrenalectomy abolished stress-induced IL-6 increases, which might point to the adrenal glands as a source of increase in circulating IL-6. This would be in line with the fact that adrenal tissue is able to produce and secrete IL-6.[96]

In addition to the large number of possible nonimmune tissue sources, which include muscle, adipose tissue, and endothelial cells, it should be noted that stress changes the number of IL-6–producing immune cells in the circulation. In some cases, this phenomenon has been shown to explain part of the stress-induced increase in NF-κB binding activity.[97] In summary, stress-induced increases in IL-6 are a robust finding, and increases are typically higher in adverse psychosocial conditions. Specific tissue sources are currently not well known.

Increased IL-6 concentrations in plasma or serum are also found in individuals suffering from long-term exposure to adverse psychosocial conditions, usually summarized as chronic stress (see Ref. 98 for a summary). Cross-sectional studies in wide age groups and with different samples sizes have shown that plasma concentrations of IL-6 (and in many cases, also CRP) are increased in individuals reporting more chronic stress, depression, or cynical distrust,[99] and are higher in unemployed persons.[100] IL-6 was repeatedly found to be increased in caregivers to relatives suffering from Alzheimer's disease, for example.[101–103] A longitudinal study further showed stronger IL-6 increases over a 6-year period in elderly caregivers compared with controls,[104] which indicates that chronic stress might be driving faster age-related increases in inflammatory mediators.

While acute stress studies are mostly relevant for understanding the pathways between CNS perception of stress and peripheral inflammation, IL-6 increases related to chronic stress, and due to the strong association of inflammatory mediators with human morbidity and mortality, are highly relevant for health, and will therefore need to be understood and addressed by targeted intervention strategies.

Summary and conclusions

The results summarized previously provide evidence that IL-6 is an important mediator of CNS-immune interaction. IL-6 secretion is enhanced in inflammatory disease such as RA, but also in stress or disordered sleep. The pleiotropic effects of IL-6 on the CNS explain why, for example, inflammatory disease is often associated with fatigue, sleepiness, or reduced sleep quality. An increase of fatigue, sleepiness, and sleep duration can be encountered in both acute and chronic disease, potentially adding to the burden that disease-specific symptoms may put on affected patients. However, the enforced shift toward less activity and more rest builds the basis for recovery and convalescence, which is why doctors have always advised patients to rest and sleep in order to get well. In chronic disease, reduced activity, exhaustion, and increased need for sleep may persist but may also reflect adaptive mechanisms that are supposed to alleviate coping. Sleep loss, reduced sleep quality, insomnia, or hypersomnia predispose to immune dysfunction, which may also involve IL-6 hypersecretion and its sequelae.

Similar to acute and chronic disease, psychosocial stress has been shown to activate the inflammatory cascade, which is seen in increased concentrations of circulating biomarkers of inflammation, of which IL-6 is the most consistently documented. While acute stress-induced increases in IL-6 might be adaptive in restoring homeostasis, long-term or repeated increases are most likely related to unfavorable health outcomes. In addition to the well-documented consequences of low-grade inflammation on the cardiovascular system, for example, the literature summarized here further shows that stress-induced IL-6 increases are closely linked to fatigue and reduced sleep quality. Since sleep is important also for recovery from psychological stress, long-term or repeated stress-induced activation of inflammation is a maladaptive response.

For inflammatory conditions in particular, these findings add important insights to the well-understood effects of anemia and nighttime pain through inflammation, which is also mediated by IL-6. Evidence clearly suggests that changes in circulating IL-6 can cause changes in sleep quality. Conversely, reductions in sleep duration, quality, or efficiency are capable of increasing peripheral IL-6 concentrations. These results are probably related to the fact that IL-6 increase can also be observed in chronic stress, as a consequence of long-term changes in stress system activity. Taken together, these findings, from various fields of research, underscore the close relationship between IL-6 signaling with CNS processes, thereby making IL-6 a promising candidate for linking adverse CNS states with physical disease.

Conflicts of interest

Martin Aringer is on advisory boards for Roche/Chugai (Tocilizumab).

References

1. Benveniste, E.N. 1992. Inflammatory cytokines within the central nervous system: sources, function, and mechanism of action. *Am. J. Physiol.* **263:** C1–C16.
2. Cannon, J.G. 2000. Inflammatory cytokines in nonpathological states. *News Physiol. Sci.* **15:** 298–303.
3. Besedovsky, H.O. & A. del Rey. 1996. Immune-neuroendocrine interactions: facts and hypotheses. *Endocr. Rev.* **17:** 64–102.
4. Sternberg, E.M. 2001. Neuroendocrine regulation of autoimmune/inflammatory disease. *J. Endocrinol.* **169:** 429–435.
5. Nishimoto, N. & T. Kishimoto. 2006. Interleukin 6: from bench to bedside. *Nat. Clin. Pract. Rheumatol.* **2:** 619–626.
6. Yamasaki, K. *et al.* 1988. Cloning and expression of the human interleukin-6 (BSF-2/IFN beta 2) receptor. *Science* **241:** 825–828.
7. Taga, T. *et al.* 1989. Interleukin-6 triggers the association of its receptor with a possible signal transducer, gp130. *Cell* **58:** 573–581.

8. Yin, T. *et al.* 1993. Involvement of IL-6 signal transducer gp130 in IL-11-mediated signal transduction. *J. Immunol.* **151:** 2555–2561.

9. Zhong, Z., Z. Wen & J.E.J. Darnell. 1994. Stat3: a STAT family member activated by tyrosine phosphorylation in response to epidermal growth factor and interleukin-6. *Science* **264:** 95–98.

10. Narazaki, M. *et al.* 1994. Activation of JAK2 kinase mediated by the interleukin 6 signal transducer gp130. *Proc. Natl. Acad. Sci. U.S.A.* **91:** 2285–2289.

11. Naka, T., N. Nishimoto & T. Kishimoto. 2002. The paradigm of IL-6: from basic science to medicine. *Arthritis Res.* **4** Suppl 3: S233–S242.

12. Rose-John, S. *et al.* 2006. Interleukin-6 biology is coordinated by membrane-bound and soluble receptors: role in inflammation and cancer. *J. Leukoc. Biol.* **80:** 227–236.

13. Papanicolaou, D.A. & A.N. Vgontzas. 2000. Interleukin-6: the endocrine cytokine. *J. Clin. Endocrinol. Metab.* **85:** 1331–1333.

14. Mohamed-Ali, V. *et al.* 2001. beta-Adrenergic regulation of IL-6 release from adipose tissue: *in vivo* and *in vitro* studies. *J. Clin. Endocrinol. Metab.* **86:** 5864–5869.

15. Sironi, M. *et al.* 1989. IL-1 stimulates IL-6 production in endothelial cells. *J. Immunol.* **142:** 549–553.

16. Ershler, W.B. 1993. Interleukin-6: a cytokine for gerontologists. *J. Am. Geriatr. Soc.* **41:** 176–181.

17. Ershler, W.B., W.H. Sun & N. Binkley. 1994. The role of interleukin-6 in certain age-related diseases. *Drugs Aging* **5:** 358–365.

18. Bruunsgaard, H. *et al.* 2003. Predicting death from tumour necrosis factor-alpha and interleukin-6 in 80-year-old people. *Clin. Exp. Immunol.* **132:** 24–31.

19. Baumann, H. & J. Gauldie. 1994. The acute phase response. *Immunol. Today* **15:** 74–80.

20. Olivadoti, M.D. & M.R. Opp. 2008. Effects of i.c.v. administration of interleukin-1 on sleep and body temperature of interleukin-6-deficient mice. *Neuroscience* **153:** 338–348.

21. Garlanda, C. *et al.* 2005. Pentraxins at the crossroads between innate immunity, inflammation, matrix deposition, and female fertility. *Annu. Rev. Immunol.* **23:** 337–366.

22. Xing, Z. *et al.* 1998. IL-6 is an antiinflammatory cytokine required for controlling local or systemic acute inflammatory responses. *J. Clin. Invest.* **101:** 311–320.

23. Gabay, C. & I. Kushner. 1999. Acute-phase proteins and other systemic responses to inflammation. *N. Engl. J. Med.* **340:** 448–454.

24. Aletaha, D. *et al.* 2010. 2010 rheumatoid arthritis classification criteria: an American College of Rheumatology/European League Against Rheumatism collaborative initiative. *Ann. Rheum. Dis.* **69:** 1580–1588.

25. Radovits, B.J. *et al.* 2010. Excess mortality emerges after 10 years in an inception cohort of early rheumatoid arthritis. *Arthritis Care Res.* **62:** 362–370.

26. Felson, D.T. *et al.* 2011. American College of Rheumatology/European League Against Rheumatism provisional definition of remission in rheumatoid arthritis for clinical trials. *Arthritis Rheum.* **63:** 573–586.

27. Szalai, A.J. *et al.* 1998. Testosterone and IL-6 requirements for human C-reactive protein gene expression in transgenic mice. *J. Immunol.* **160:** 5294–5299.

28. Genovese, M.C. *et al.* 2008. Interleukin-6 receptor inhibition with tocilizumab reduces disease activity in rheumatoid arthritis with inadequate response to disease-modifying antirheumatic drugs: the tocilizumab in combination with traditional disease-modifying antirheumatic drug therapy study. *Arthritis Rheum.* **58:** 2968–2980.

29. Nishimoto, N. *et al.* 2007. Study of active controlled monotherapy used for rheumatoid arthritis, an IL-6 inhibitor (SAMURAI): evidence of clinical and radiographic benefit from an x ray reader-blinded randomised controlled trial of tocilizumab. *Ann. Rheum. Dis.* **66:** 1162–1167.

30. Smolen, J.S. *et al.* 2008. Effect of interleukin-6 receptor inhibition with tocilizumab in patients with rheumatoid arthritis (OPTION study): a double-blind, placebo-controlled, randomised trial. *Lancet* **371:** 987–997.

31. Kremer, J.M. *et al.* 2011. Tocilizumab inhibits structural joint damage in rheumatoid arthritis patients with inadequate responses to methotrexate: results from the double-blind treatment phase of a randomized placebo-controlled trial of tocilizumab safety and prevention of structural joint damage at one year. *Arthritis Rheum.* **63:** 609–621.

32. Burmester, G.R. *et al.* 2011. Effectiveness and safety of the interleukin 6-receptor antagonist tocilizumab after 4 and 24 weeks in patients with active rheumatoid arthritis: the first phase IIIb real-life study (TAMARA). *Ann. Rheum. Dis.* **70:** 755–759.

33. Iking-Konert, C. *et al.* 2011. Performance of the new 2011 ACR/EULAR remission criteria with tocilizumab using the phase IIIb study TAMARA as an example and their comparison with traditional remission criteria. *Ann. Rheum. Dis.* **70:** 1986–1990.

34. Han, C. *et al.* 2007. Association of anemia and physical disability among patients with rheumatoid arthritis. *J. Rheumatol.* **34:** 2177–2182.

35. Nikolaisen, C., Y. Figenschau & J.C. Nossent. 2008. Anemia in early rheumatoid arthritis is associated with interleukin 6-mediated bone marrow suppression, but has no effect on disease course or mortality. *J. Rheumatol.* **35:** 380–386.

36. Emery, P. *et al.* 2008. IL-6 receptor inhibition with tocilizumab improves treatment outcomes in patients with rheumatoid arthritis refractory to anti-tumour necrosis factor biologicals: results from a 24-week multicentre randomised placebo-controlled trial. *Ann. Rheum. Dis.* **67:** 1516–1523.

37. Nemeth, E. *et al.* 2004. Hepcidin regulates cellular iron efflux by binding to ferroportin and inducing its internalization. *Science* **306:** 2090–2093.

38. Nemeth, E. *et al.* 2004. IL-6 mediates hypoferremia of inflammation by inducing the synthesis of the iron regulatory hormone hepcidin. *J. Clin. Invest.* **113:** 1271–1276.

39. Hess, A. *et al.* 2011. Blockade of TNF-alpha rapidly inhibits pain responses in the central nervous system. *Proc. Natl. Acad. Sci. U.S.A.* **108:** 3731–3736.

40. Hewlett, S. *et al.* 2011. Fatigue in rheumatoid arthritis: time for a conceptual model. *Rheumatology* **50:** 1004–1006.

41. Borbely, A.A. 1982. A two process model of sleep regulation. *Hum. Neurobiol.* **1:** 195–204.

42. Morris, C.J., D. Aeschbach & F.A. Scheer. 2012. Circadian system, sleep and endocrinology. *Mol. Cell Endocrinol.* **349:** 91–104.

43. Neu, D. *et al.* 2010. Do 'sleepy' and 'tired' go together? Rasch analysis of the relationships between sleepiness, fatigue and nonrestorative sleep complaints in a nonclinical population sample. *Neuroepidemiology* **35:** 1–11.

44. Gangwisch, J.E. *et al.* 2006. Short sleep duration as a risk factor for hypertension: analyses of the first National Health and Nutrition Examination Survey. *Hypertension* **47:** 833–839.

45. Gangwisch, J.E. *et al.* 2007. Sleep duration as a risk factor for diabetes incidence in a large U.S. sample. *Sleep* **30:** 1667–1673.

46. Grandner, M.A. *et al.* 2010. Mortality associated with short sleep duration: the evidence, the possible mechanisms, and the future. *Sleep Med. Rev.* **14:** 191–203.

47. Patel, S.R. *et al.* 2004. A prospective study of sleep duration and mortality risk in women. *Sleep* **27:** 440–444.

48. Ancoli-Israel, S. 2006. The impact and prevalence of chronic insomnia and other sleep disturbances associated with chronic illness. *Am. J. Manag. Care* **12:** S221–S229.

49. Irwin, M. 2002. Effects of sleep and sleep loss on immunity and cytokines. *Brain Behav. Immun.* **16:** 503–512.

50. Okun, M.L., M. Coussons-Read & M. Hall. 2009. Disturbed sleep is associated with increased C-reactive protein in young women. *Brain Behav. Immun.* **23:** 351–354.

51. Patel, S.R. *et al.* 2009. Sleep duration and biomarkers of inflammation. *Sleep* **32:** 200–204.

52. Vgontzas, A.N. *et al.* 1997. Elevation of plasma cytokines in disorders of excessive daytime sleepiness: role of sleep disturbance and obesity. *J. Clin. Endocrinol. Metab.* **82:** 1313–1316.

53. Vgontzas, A.N. *et al.* 1999. Circadian interleukin-6 secretion and quantity and depth of sleep. *J. Clin. Endocrinol. Metab.* **84:** 2603–2607.

54. Vgontzas, A.N. *et al.* 2005. IL-6 and its circadian secretion in humans. *Neuroimmunomodulation* **12:** 131–140.

55. Redwine, L. *et al.* 2000. Effects of sleep and sleep deprivation on interleukin-6, growth hormone, cortisol, and melatonin levels in humans. *J. Clin. Endocrinol. Metab.* **85:** 3597–3603.

56. Dimitrov, S. *et al.* 2006. Sleep enhances IL-6 trans-signaling in humans. *FASEB J.* **20:** 2174–2176.

57. Motzkus, D., U. Albrecht & E. Maronde. 2002. The human PER1 gene is inducible by interleukin-6. *J. Mol. Neurosci.* **18:** 105–109.

58. Cavadini, G. *et al.* 2007. TNF-alpha suppresses the expression of clock genes by interfering with E-box–mediated transcription. *Proc. Natl. Acad. Sci. U.S.A.* **104:** 12843–12848.

59. Vgontzas, A.N. *et al.* 2004. Adverse effects of modest sleep restriction on sleepiness, performance, and inflammatory cytokines. *J. Clin. Endocrinol. Metab.* **89:** 2119–2126.

60. Vgontzas, A.N. *et al.* 2007. Daytime napping after a night of sleep loss decreases sleepiness, improves performance, and causes beneficial changes in cortisol and interleukin-6 secretion. *Am. J. Physiol. Endocrinol. Metab.* **292:** E253–E261.

61. Shearer, W.T. *et al.* 2001. Soluble TNF-alpha receptor 1 and IL-6 plasma levels in humans subjected to the sleep deprivation model of spaceflight. *J. Allergy Clin. Immunol.* **107:** 165–170.

62. Hogan, D. *et al.* 2003. Interleukin-6 alters sleep of rats. *J. Neuroimmunol.* **137:** 59–66.

63. Krueger, J.M. & J.A. Majde. 2003. Humoral links between sleep and the immune system: research issues. *Ann. N.Y. Acad. Sci.* **992:** 9–20.

64. Obal, F.J. *et al.* 1990. Interleukin 1 alpha and an interleukin 1 beta fragment are somnogenic. *Am. J. Physiol.* **259:** R439–R446.

65. Opp, M.R., F.J. Obal & J.M. Krueger. 1991. Interleukin 1 alters rat sleep: temporal and dose-related effects. *Am. J. Physiol.* **260:** R52–R58.

66. Spath-Schwalbe, E. *et al.* 1998. Acute effects of recombinant human interleukin-6 on endocrine and central nervous sleep functions in healthy men. *J. Clin. Endocrinol. Metab.* **83:** 1573–1579.

67. Crofford, L.J. *et al.* 1997. Circadian relationships between interleukin (IL)-6 and hypothalamic-pituitary-adrenal axis hormones: failure of IL-6 to cause sustained hypercortisolism in patients with early untreated rheumatoid arthritis. *J. Clin. Endocrinol. Metab.* **82:** 1279–1283.

68. Okun, M.L. *et al.* 2004. Exploring the cytokine and endocrine involvement in narcolepsy. *Brain Behav. Immun.* **18:** 326–332.

69. Tauman, R., L.M. O'Brien & D. Gozal. 2007. Hypoxemia and obesity modulate plasma C-reactive protein and interleukin-6 levels in sleep-disordered breathing. *Sleep Breath* **11:** 77–84.

70. Teramoto, S., H. Yamamoto & Y. Ouchi. 2003. Increased C-reactive protein and increased plasma interleukin-6 may synergistically affect the progression of coronary atherosclerosis in obstructive sleep apnea syndrome. *Circulation* **107:** E40–E50.

71. Teramoto, S., H. Yamamoto & Y. Ouchi. 2004. Increased plasma interleukin-6 is associated with the pathogenesis of obstructive sleep apnea syndrome. *Chest* **125:** 1964–65; author reply, 1965.

72. Vgontzas, A.N. *et al.* 2000. Chronic systemic inflammation in overweight and obese adults. *JAMA* **283:** 2235.

73. Mehra, R. *et al.* 2006. Soluble interleukin 6 receptor: a novel marker of moderate to severe sleep-related breathing disorder. *Arch. Intern. Med.* **166:** 1725–1731.

74. Fragiadaki, K. *et al.* 2012. Sleep disturbances and interleukin 6 receptor inhibition in rheumatoid arthritis. *J. Rheumatol.* **39:** 60–62.

75. Chauffier, K. *et al.* 2012. Effect of biotherapies on fatigue in rheumatoid arthritis: a systematic review of the literature and meta-analysis. *Rheumatology* **51:** 60–68.

76. Papanicolaou, D.A. *et al.* 1996. Exercise stimulates interleukin-6 secretion: inhibition by glucocorticoids and correlation with catecholamines. *Am. J. Physiol.* **271:** E601–E605.

77. Petersen, A.M. & B.K. Pedersen. 2005. The anti-inflammatory effect of exercise. *J. Appl. Physiol.* **98:** 1154–1162.

78. LeMay, L.G., A.J. Vander & M.J. Kluger. 1990. The effects of psychological stress on plasma interleukin-6 activity in rats. *Physiol. Behav.* **47:** 957–961.

79. Zhou, D. *et al.* 1993. Exposure to physical and psychological stressors elevates plasma interleukin 6: relationship to the activation of hypothalamic-pituitary-adrenal axis. *Endocrinology* **133:** 2523–2530.

80. Steptoe, A., M. Hamer & Y. Chida. 2007. The effects of acute psychological stress on circulating inflammatory factors in humans: a review and meta-analysis. *Brain Behav. Immun.* **21:** 901–912.

81. Brydon, L. *et al.* 2004. Socioeconomic status and stress-induced increases in interleukin-6. *Brain Behav. Immun.* **18:** 281–290.

82. Hamer, M. & A. Steptoe. 2007. Association between physical fitness, parasympathetic control, and proinflammatory responses to mental stress. *Psychosom. Med.* **69:** 660–666.

83. Pace, T.W. *et al.* 2009. Effect of compassion meditation on neuroendocrine, innate immune and behavioral responses to psychosocial stress. *Psychoneuroendocrinology* **34:** 87–98.

84. Carroll, J.E. *et al.* 2011. Negative affective responses to a speech task predict changes in interleukin (IL)-6. *Brain Behav. Immun.* **25:** 232–238.

85. O'Donnell, K. *et al.* 2008. Self-esteem levels and cardiovascular and inflammatory responses to acute stress. *Brain Behav. Immun.* **22:** 1241–1247.

86. Kiecolt-Glaser, J.K. *et al.* 2010. Stress, inflammation, and yoga practice. *Psychosom. Med.* **72:** 113–121.

87. Brydon, L. *et al.* 2009. Synergistic effects of psychological and immune stressors on inflammatory cytokine and sickness responses in humans. *Brain Behav. Immun.* **23:** 217–224.

88. Carpenter, L.L. *et al.* 2010. Association between plasma IL-6 response to acute stress and early-life adversity in healthy adults. *Neuropsychopharmacology* **35:** 2617–2623.

89. Pace, T.W. *et al.* 2006. Increased stress-induced inflammatory responses in male patients with major depression and increased early life stress. *Am. J. Psychiatry* **163:** 1630–1633.

90. Bierhaus, A. *et al.* 2003. A mechanism converting psychosocial stress into mononuclear cell activation. *Proc. Natl. Acad. Sci. U.S.A.* **100:** 1920–1925.

91. Wolf, J.M. *et al.* 2009. Determinants of the NF-kappaB response to acute psychosocial stress in humans. *Brain Behav. Immun.* **23:** 742–749.

92. Sondergaard, S.R. *et al.* 2000. Changes in plasma concentrations of interleukin-6 and interleukin-1 receptor antagonists

93. van Gool, J. *et al.* 1990. The relation among stress, adrenalin, interleukin 6 and acute phase proteins in the rat. *Clin. Immunol. Immunopathol.* **57:** 200–210.

94. DeRijk, R.H. *et al.* 1994. Induction of plasma interleukin-6 by circulating adrenaline in the rat. *Psychoneuroendocrinology* **19:** 155–163.

95. von Kanel, R. *et al.* 2008. Aspirin, but not propranolol, attenuates the acute stress-induced increase in circulating levels of interleukin-6: a randomized, double-blind, placebo-controlled study. *Brain Behav. Immun.* **22:** 150–157.

96. Judd, A.M. 1998. Cytokine expression in the rat adrenal cortex. *Horm. Metab. Res.* **30:** 404–410.

97. Richlin, V.A. *et al.* 2004. Stress-induced enhancement of NF-kappaB DNA-binding in the peripheral blood leukocyte pool: effects of lymphocyte redistribution. *Brain Behav. Immun.* **18:** 231–237.

98. Hansel, A. *et al.* 2010. Inflammation as a psychophysiological biomarker in chronic psychosocial stress. *Neurosci. Biobehav. Rev.* **35:** 115–121.

99. Ranjit, N. *et al.* 2007. Psychosocial factors and inflammation in the multi-ethnic study of atherosclerosis. *Arch. Intern. Med.* **167:** 174–181.

100. Hintikka, J. *et al.* 2009. Unemployment and ill health: a connection through inflammation? *BMC Public Health* **9:** 410.

101. Lutgendorf, S.K. *et al.* 1999. Life stress, mood disturbance, and elevated interleukin-6 in healthy older women. *J. Gerontol. A Biol. Sci. Med. Sci.* **54:** M434–M439.

102. von Känel, R. *et al.* 2006. Effect of Alzheimer caregiving stress and age on frailty markers interleukin-6, C-reactive protein, and D-dimer. *J. Gerontol. A Biol. Sci. Med. Sci.* **61:** 963–969.

103. Clark, M.C. *et al.* In press. Psychosocial and biological indicators of depression in the care giving population. *Biol. Res. Nurs.*

104. Kiecolt-Glaser, J.K. *et al.* 2003. Chronic stress and age-related increases in the proinflammatory cytokine IL-6. *Proc. Natl. Acad. Sci. U.S.A.* **100:** 9090–9095.

Ann. N.Y. Acad. Sci. ISSN 0077-8923

ANNALS OF THE NEW YORK ACADEMY OF SCIENCES

Issue: *Neuroimmunomodulation in Health and Disease*

Role of sleep in the regulation of the immune system and the pituitary hormones

Beatriz Gómez-González,* Emilio Domínguez-Salazar,* Gabriela Hurtado-Alvarado, Enrique Esqueda-Leon, Rafael Santana-Miranda, Jose Angel Rojas-Zamorano, and Javier Velázquez-Moctezuma

Department of Biology of Reproduction and Sleep Disorders Clinic, Universidad Autónoma Metropolitana-Iztapalapa, Mexico City, Federal District, Mexico

Address for correspondence: Javier Velázquez-Moctezuma, M.D., PhD., Area of Neurosciences, Department of Biology of Reproduction, CBS, Universidad Autónoma Metropolitana-Iztapalapa, Av. San Rafael Atlixco No. 186, Col Vicentina, Iztapalapa, Mexico City, DF, Mexico 09340. jvm@xanum.uam.mx

Sleep is characterized by a reduced response to external stimuli and a particular form of electroencephalographic (EEG) activity. Sleep is divided into two stages: REM sleep, characterized by muscle atonia, rapid eye movements, and EEG activity similar to wakefulness, and non-REM sleep, characterized by slow EEG activity. Around 80% of total sleep time is non-REM. Although it has been intensely studied for decades, the function (or functions) of sleep remains elusive. Sleep is a highly regulated state; some brain regions and several hormones and cytokines participate in sleep regulation. This mini-review focuses on how pituitary hormones and cytokines regulate or affect sleep and how sleep modifies the plasma concentration of hormones as well as cytokines. Also, we review the effects of hypophysectomy and some autoimmune diseases on sleep pattern. Finally, we propose that one of the functions of sleep is to maintain the integrity of the neuro–immune–endocrine system.

Keywords: sleep; cytokines; pituitary gland; hormones; autoimmune diseases; sleep loss; sleep function

Introduction

Sleep is a widespread phenomenon among vertebrates; it is characterized by the adoption of a species-specific posture, by increases in sensory thresholds, immobility, and by the generation of hallmark patterns of electroencephalographic activity.[1] In mammals, sleep is divided into two stages: rapid eye movement (REM) sleep and non-REM sleep. During REM sleep, a high-frequency and low-voltage electrical brain activity is observed, accompanied by rapid eye movements and absence of muscle tone in skeletal muscles. On the other hand, non-REM sleep, composed of both light and slow-wave sleep, is characterized by the presence of high-voltage and low-frequency electroencephalographic activity, diminished muscle tone, and slow

eye movements.[2,3] Humans sleep around one-third of their lifetime, and other mammals, such as rodents, sleep almost two-thirds of their lifetime.

Sleep is an active process that requires the synchronous activation and deactivation of many nervous centers located in the hypothalamus and brain stem, whose activity directly modifies the cortical electrical activity. At the central nervous system level, REM sleep is characterized by a decreased release of noradrenalin and serotonin and by increases in acetylcholine release by the REM-sleep generators in the pons; in the hypothalamus, histaminergic and orexinergic neurons are silent during REM sleep compared with wakefulness.[4] Meanwhile, non-REM sleep is characterized by almost full suppression of acetylcholine release and a progressive reduction in noradrenalin and serotonin concentration.[5] In addition to the changes observed in neurotransmitter content, a tight reciprocal relationship between sleep and the endocrine system

---

*These authors contributed equally to the work reported.

doi: 10.1111/j.1749-6632.2012.06616.x
Ann. N.Y. Acad. Sci. 1261 (2012) 97–106 © 2012 New York Academy of Sciences.

was described long ago in both human and animal models. Anterior and posterior pituitary hormone secretion has been shown to occur during specific sleep stages in humans and, reciprocally, sleep loss modifies their plasma concentration.[6] Recently, sleep has also been shown to influence cytokine plasma levels and to be regulated by those immune molecules during both normal and pathological conditions.[7]

Pituitary hormonal secretion during sleep

Since the pioneer work of Sassin *et al.*,[8] it has been recognized that the sleep/wake cycle is important for hormone secretion; the release pattern of almost every hormone is influenced by a circadian rhythm, and pituitary hormones are not the exception. Adrenocorticotropic hormone (ACTH) is released in pulses over the whole day; however, it reaches its zenith before awakening.[9] Similarly, follicle-stimulating hormone (FSH) pulses occur about every 2 h, and luteinizing hormone (LH) release occurs above every 30–90 min; nevertheless, the highest plasma concentration of FSH takes place before awakening, and the zenith of LH occurs in the middle of sleep time.[10,11] The plasma concentration of TSH and prolactin also rises at the beginning of sleep, reaching its maximum concentration 2 h later and decreasing 2 h before awakening.[12,13] A similar release pattern is observed for growth hormone (GH), which starts to be released at the beginning of the sleep period, reaching its maximum concentration 2 h later. Four hours after sleep onset, the concentration of GH is similar to that of wakefulness.[13] Postmenopausal women present a reduced GH peak at night and a higher extra peak in the afternoon; however, hormonal therapy restores GH secretion to normal premenopausal patterns.[14] Studies show that lower sleep-onset release of GH and prolactin in older people, mainly women, is related to lower levels of delta activity during non-REM sleep, suggesting that non-REM sleep is responsible for inducing the release of those hormones.[15] Therefore, sleep maintains a tight reciprocal relationship with anterior pituitary hormone release.

Posterior pituitary hormones are also influenced by the sleep/wake rhythm. Plasma arginine vasopressin (AVP) concentration normally rises at night; this increase is suppressed in patients with nocturia, particularly in older people.[16] Nighttime levels of oxytocin, measured in pregnant women, were higher than daytime levels; moreover, the further along in the pregnancy, the higher the level of plasma oxytocin.[17] These results indicate that all the pituitary hormones are released in higher concentration during sleep, suggesting both that sleep is responsible to keep the integrity of the endocrine system, and that the pituitary hormones are able to induce sleep.

Effect of administration of pituitary hormones on sleep

Posterior pituitary hormone administration has been shown to induce contrasting effects on sleep pattern; in humans, the administration of desmopressin, a synthetic analog of AVP used for the treatment of nocturnal enuresis, induced dizziness and daytime somnolence.[18] Likewise, in elderly humans, the subchronic intranasal administration of AVP (for 30 days) increased slow wave sleep,[19] and the intravenous administration of AVP in young adult men decreased REM sleep.[20] In contrast, AVP administration for 1 week in children did not affect the sleep pattern.[21] Vasotocin, an oligopeptide hybrid between oxytocin and AVP, present in nonmammal vertebrates and fetal mammals,[22] increased the amount of REM sleep, decreased REM sleep latency, and induced REM sleep episodes at sleep onset when intranasally administered to healthy prepubertal boys; however, neither AVP nor oxytocin were able to induce those effects in the same participants.[23] In other mammals, such as the rat, intracerebroventricular administration of oxytocin increased the percentage of wakefulness and decreased the percentage of non-REM and REM sleep.[24]

With respect to anterior pituitary hormones, the administration of several doses of GH increased REM sleep in the rat.[25] In humans, GH deficiency syndrome is characterized by a reduction in sleep, and the administration of recombinant human GH generally helps to recover sleep;[26] however, in about 25% of patients, GH treatment induced sleep distress.[27] On the other hand, the administration of prolactin,[28] the induction of the release of endogenous prolactin,[29] and chronic hyperprolactinemia[30] are also associated with increased REM sleep. Moreover, it has been suggested that the increase of REM sleep after acute stress in animal models, for example, by immobilization or ether brief exposure, is induced by an increase in prolactin plasma concentration.[31]

Altered sleep patterns are also associated with go-nadotrophins and thyrotropin; in narcolepsy, a disease characterized by excessive daytime sleepiness, sleep fragmentation at nighttime, cataplexy, REM sleep intrusion into wakefulness, and decreased levels of LH have been described.[32] In contrast, the other gonadotrophin FSH, given to idiopathic hypogonadotropic hypogonadism patients, reduces the excessively high levels of slow-wave sleep.[33] In infant patients with hypothyroidism, slow-wave sleep is reduced, and active sleep (an early form of REM sleep) is increased in comparison with healthy same-age controls.[34] However, data on adults show that TSH appears to be released during REM sleep, and that TSH or thyroxin administration is capable of regulating the sleep/wake pattern in older men and in subjects with depression or hypothyroidism.[35]

Because of their release pattern, with a zenith near waking time, hormones of the hypothalamus–pituitary–adrenal (HPA) axis are generally considered as wakefulness inducers;[36] however, acute stressors are able to induce an increase of REM sleep in the following sleep episode.[37] Some evidence shows that high doses of corticosterone increase waking time and decrease non-REM sleep, however, low doses of glucocorticoids induce the opposite effect, a slight increase in non-REM sleep.[37]

Although corticosterone is the main effector of the stress response, its administration does not increase REM sleep, such as the reported increase after immobilization or electric foot shock stress.[38] Therefore, the stress-dependent increase in REM sleep has been attributed to corticotrophin-releasing hormone (CRH) in addition to prolactin; CRH can be released without the induction of the release of ACTH; in that case, CRH increases REM sleep. When ACTH is released by CRH, the effect of CRH on REM sleep is not observed because it predominates the effect of ACTH wakefulness inducer.[36] Figure 1 shows the effect of some pituitary hormones on non-REM and REM sleep in humans.

Effects of hypophysectomy on the sleep pattern in both human and animal models

Loss of the pituitary with concomitant changes in plasma concentrations of pituitary hormones is one common finding in clinical practice; pituitary tumors are the third most common brain tumor after meningioma and glioblastoma (the population rate may be as high as 1.05 per 100,000).[39] The most common treatment for pituitary tumors is surgical resection, which implies total or partial hypophysectomy with a subsequent hormonal reduction in plasma concentrations.[40] Although experimental

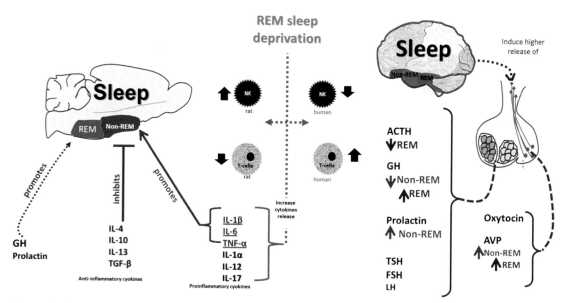

Figure 1. Sleep as a regulator of the neuro–immune–endocrine interaction. Image shows the reciprocal relationship between sleep and the immune system (cytokines and cells), as well as between sleep and the endocrine system (pituitary hormones), in physiological conditions and sleep deprivation in both the human and the rodent animal model.

data obtained in animal models long ago demonstrated a close relationship between sleep and pituitary hormones,[31,41,42] the sleep consequences of pituitary hypofunctioning, secondary to partial hypophysectomy in humans, have only recently been described.[43] Patients subjected to pituitary surgical resection with diminished plasma levels of GH, LH, FSH, ACTH, and TSH, presented a reduction in sleep efficiency, mainly due to a dramatic reduction in REM sleep; they also presented increased duration in light sleep (stages 1 and 2 of non-REM sleep) and increased waking time at night.[43]

Pioneer studies carried out by Jouvet's group in the cat showed that total hypophysectomy, the surgical removal of both anterior and posterior pituitary lobes, did not induce systematic changes in the sleep pattern during the first 15 days post hypophysectomy under basal conditions nor under sleep deprivation conditions, compared with intact control animals.[41] However, the same group also reported that in the rat, total hypophysectomy did modify the sleep pattern in the short term.[42] Zhang *et al.*[42] showed that total hypophysectomized rats presented reduced duration of REM sleep under basal conditions, compared with intact controls; moreover, when subjecting hypophysectomized rats to sleep deprivation, Zhang *et al.*[42] showed that rats whose pituitary was removed presented a REM sleep rebound of smaller magnitude and duration than that observed in intact controls. More recently, Bodosi *et al.*[31] showed that total hypophysectomy in the rat led to basal reduction of REM sleep duration; they also showed that stress exposure, a known condition that increases REM sleep duration in the following rest episode, induced a smaller REM sleep rebound during the first 4 hours poststress in the totally hypophysectomized rats compared with intact controls.[31]

Sleep regulation by the immune system: the role of cytokines

The central nervous system and the immune system maintain bidirectional communication. As a result of such interaction, if sleep is altered, there is a disruption in the response against invading microorganisms; likewise, the immune response is capable of inducing sleep changes.[44] Sleep disturbances, such as deprivation or chronic restriction, increase proinflammatory cytokines release.[45] Similarly, the increase of these chemical messengers alters brain and behavioral processes including sleep.[7] The central nervous system detects peripheral immune system activation through the stimulation of nerve fibers located at the site where the infection occurs (i.e., vagal stimulation),[44] through circulating cytokines that trespass the circumventricular organs and the blood–brain barrier,[46] and through brain-derived cytokines.[44]

Cytokines are synthesized and released in the central nervous system by both neurons and glia; neurons immunoreactive to IL-1β and TNF-α are located in brain regions involved in sleep/wake cycle regulation (e.g., hypothalamus, hippocampus, brainstem, and neocortex). In these regions, cytokine receptors are located in both neurons and astrocytes.[47–50] In mammals, it has been shown that the administration of IL-1α, IL-1β, IL-2, IL-4, IL-6, IL-8, IL-10, IL-13, IL-15, IL-18, TNF-α, TNF-β, IFN-α, IFN-β, INF-γ, or macrophage inhibitory protein 1β (also known as CCL4) modifies the sleep pattern. Particularly, IL-1β, IL-2, IL-6, IL-18, and TNF-α (all proinflammatory cytokines) promote non-REM sleep, whereas IL-4, IL-10, IL-13, and TGF-β (all anti-inflammatory cytokines) inhibit non-REM sleep.[51] Furthermore, the cytokine-dependent effect on the sleep pattern may be reverted by the coadministration of specific antibodies against those cytokines, by soluble receptors or by antagonists.[52] The hypothesized mechanism includes changes in the action and secretion of neurotransmitters (such as monoamines, acetylcholine, and glutamate), of hormones (CRH, ACTH, MSH, and GH), and changes in synaptic plasticity at neural circuits regulating mood, motor activity, motivation, anxiety, and alertness.[53]

Under physiological conditions, IL-1β, IL-6, and TNF-α exhibit a circadian rhythm, with a maximum peak during the night and a nadir during wakefulness.[54] In patients with insomnia, narcolepsy, and obstructive sleep apnea hypopnea (OSAH), an increase in the level of serum TNF-α has been observed. Interestingly, CPAP treatment for OSAH decreases serum TNF-α.[55] Another finding indicating a role of TNF-α in sleep is observed in patients with a large number of awakenings during the night, who present with increased levels of TNF-α soluble receptor.[56] In the case of IL-6, it has been shown that prolonged sleep restriction, by breathing disorders or voluntary sleep restriction, increases IL-6 serum levels.[55,57,58]

Sleep loss and the immune response

Cellular immunity

Several studies have reported controversial effects of sleep deprivation on the number of circulating white blood cells in human volunteers.[59–62] Irwin *et al.*[59] reported increased numbers of lymphocytes and decreased numbers of neutrophils and NK cells in the early morning after partial sleep restriction for 1 night (subjects were allowed to sleep ~3.5 h) and, after 1 night of sleep recovery, the white cell count attained basal levels. Likely, Heiser *et al.*[60] found that total sleep deprivation for ~36 h reduced morning NK cell numbers but increased monocytes and T helper cell counts in peripheral blood; however, sleep recovery promoted a further reduction in the number of NK cells and a return to basal levels of T helper cells. Born *et al.*[61] found that total sleep deprivation for 36 h abolished NK cell circadian rhythm during both the deprivation day and the following day; meanwhile, increased numbers of lymphocytes (both B cells and T helper) at nighttime were observed only during the deprivation day. More recently, Ruiz *et al.*[62] described an increase in the number of neutrophils and T lymphocytes after 2 days of total sleep deprivation or 4 days of REM-sleep deprivation in humans; interestingly, only neutrophils reached basal values after 1 night of sleep recovery, while high numbers of T cells persisted even after 3 nights of sleep recovery. Paradoxically, in a rodent model we found that REM sleep deprivation for 10 days increased the number of NK cells, but decreased the number of T lymphocytes.[63] The reported differences among human studies in the effects of sleep loss on white blood cell counts seem to arise from the length of sleep deprivation and of the age of the participants. Despite the varying reported effects, it seems plausible to consider that sleep contributes to the regulation of circulating white blood cells, although more controlled experiments in humans are needed to elucidate the effect of sleep loss and recovery on immune cell counts and functions.

Humoral immunity

Immunological signaling molecules, like cytokines and chemokines, are important in brain–immune system communication. Sleep loss generally increases the production of proinflammatory molecules. In animal models, REM sleep deprivation increased circulating levels of IL-6, IL-12, IL-1α, IL-1β, IL-17, and TNF-α,[45,64] and, at least for IL-17 and TNF-α the higher levels persisted even 1 week after the end of the REM sleep deprivation period.[45] Moreover, it is known that acute sleep restriction, in both human and animal models, lowers immune response as is shown by decreases in antibody titers after influenza and hepatitis A vaccination.[65–67]

Sleep deprivation and infection

In animal models, chronic sleep deprivation (both total and selectively for REM sleep) has a lethal effect.[68] Several reports have shown that sleep-deprived animals have body weight loss in spite of increased food intake;[69] they also present hyperthermia[70] and chronic deterioration of health, apparently derived from generalized infections without exposure to external pathogens.[68,71] After prolonged sleep deprivation, bacteremia develops in a rodent model; Everson and Toth[71] found bacteria presumably translocated from the intestine to the blood and several extraintestinal sites, such as the lymph nodes, lungs, and liver of sleep-deprived rats, compared with control animals sleeping *ad libitum*. Therefore, the lethal effect of sleep deprivation could be related to both increased metabolic rate and decreased host defense against pathogens.

Sleep deprivation and microglia

Microglia are the first line of defense in the brain; they exert phagocyte functions, constitute the only antigen-presenting cell, and release numerous immune mediators, for example, cytokines and chemokines, into the central nervous system. Resting microglia phenotype is characterized by long, thin, and spiny ramifications; after central nervous system damage, microglia become hyperramified and progressively acquire a reactive phagocytic form.[72] In animal models, it has been shown that brief total sleep deprivation (of only 1–2 h) may reduce the expression of reactive microglia markers (e.g., CD11b),[73] while prolonged total sleep deprivation (for 5 days) increases the number of hyperramified and reactive microglia and astroglia in the rat hippocampus compared with rats sleeping normally;[74,75] Moreover, gliosis secondary to sleep deprivation may serve as a preconditioning stimulus against central nervous system insults, such as ischemia.[75] Hsu *et al.*[75] found that 5-day sleep deprivation before brain ischemia protected the hippocampus from extensive neuronal cell death (at least in CA1) and reduced gliosis (both microglial and astroglial) in the postischemia period. Weil *et al.*[76] also showed that the postischemia decreased neuronal death after acute REM sleep deprivation was related to decreases in IL-1β brain concentration and increases in the brain concentration of the antiinflammatory cytokine IL-10.

Autoimmune diseases and sleep

Sleep disturbances are common in many medical disorders, and the therapies to treat them may also disrupt sleep. Autoimmune diseases are medical conditions characterized by immune attack against the organism's own cells. Common sleep symptomatology in patients with autoimmune diseases includes sleep disruption and excessive daytime sleepiness. The main polysomnographic findings in autoimmune disease patients are shown in Table 1. Moreover, autoimmune diseases increase the risk of developing other sleep disorders, as shown in Table 1. The observed sleep fragmentation, sleep breathing disorders, or sleep-related movement disorders may exert detrimental effects on cognitive skills and mood of the patients.

Table 1. Polysomnographic findings and sleep disorders associated in patients with autoimmune diseases[a]

Autoimmune disease	Wake	Arousal	REM	TST	WASO	SL	S1	S2	SWS	SE	Associated sleep disorders
Myasthenia gravis[77,78]	↑	↑	↓	↓	N	↑	↑	=	↓	↓	OSA[77,78]
Juvenile rheumatoid arthritis[79–81]	N	↑	=	↑	↑	=	=	=	↓	=	OSA,[81] parasomnias[81,91]
Rheumatoid arthritis[82]	N	↑	N	=	↑	=	=	=	N	=	OSA,[92] RLS/PLMS[93]
Systemic lupus erythematosus[83,84]	=	↑	N	N	N	=	↑	N	↓	↓	OSA,[83,84] RLS/PLMS,[83,84] narcolepsy with cataplexy[84]
Ankylosing spondylitis[85]	N	N	N	N	N	N	↑	N	↑	N	OSA[94]
Sjogren's syndrome[86]	↑	↑	N	↓	↑	N	N	N	N	↓	RLS/PLMS[86]
Scleroderma[87]	N	↑	↓	N	N	N	N	N	↑	↓	RLS/PLMS[87]
Osteoarthritis[88]	N	N	N	N	N	N	↑	↓	N	N	
Type 1 diabetes[89]	N	N	=	=	=	N	=	↑	=	N	OSA[89]
Multiple sclerosis[90]	N	↑	=	N	N	=	=	=	=	↓	RLS/PLMS[99]
Bechet's disease	N	N	N	N	N	N	N	N	N	N	OSA, CSA, palatal myoclonus[95]
Sarcoidosis	N	N	N	N	N	N	N	N	N	N	OSA,[96,97] RLS/PLMS,[97] narcolepsy[98]

[a]Some autoimmune diseases are shown with their polysomnographic characteristics. The last column shows the sleep disorders associated with each autoimmune disease. REM, rapid eye movement; TST, total sleep time; WASO, wakefulness after sleep onset; SL, sleep latency; S1, stage 1; S2, stage 2; SWS, slow wave sleep, SE, sleep efficiency; ↑, increases, ↓, decreases; N, No available information; = , without change; OSA, obstructive sleep apnea; CSA, central sleep apnea; RLS, restless leg syndrome; PLMS, periodic leg movements during sleep.

Perspective: sleep is responsible for the integrity of the neuro—immune–endocrine system

Throughout this review we have shown that both hormones and cytokines are capable of modifying the sleep pattern, whereas normal sleep highly regulates the expression of several of these hormones and cytokines. Figure 1 presents this overview. Some hypotheses suggest that sleep functions include the promotion of body restoration, memory consolidation, and recovery of neurotransmitters.[100] However, all of those effects can be explained by the normal function of the endocrine and immune systems. If sleep loss is capable of modifying the timing of the release of both hormones and cytokines, then it is possible that several, if not all, the effects observed during sleep are due to the fact that sleep preserves the integrity of the neuro–immune–endocrine interaction.

Acknowledgments

This work was supported by Grant 14411150 from PROMEP-SEP and 1440712 from UAM. The authors express their gratitude to Edith Monroy for her expert review of the language of the manuscript.

Conflicts of interest

The authors declare no conflicts of interest.

References

1. Siegel, J.M. 2008. Do all animals sleep? *Trends Neurosci.* **31:** 208–213.
2. Dement, W. & N. Kleitman. 1957. Cyclic variations in EEG during sleep and their relation to eye movements, body motility, and dreaming. *Electroencephal. Clin. Neurophysiol.* **9:** 673–690.
3. Datta, S. & R.R. MacLean. 2007. Neurobiological mechanisms for the regulation of mammalian sleep–wake behavior: reinterpretation of historical evidence and inclusion of contemporary cellular and molecular evidence. *Neurosci. Biobehav. Rev.* **31:** 775–824.
4. Reinoso-Suárez, F., I. de Andrés, M.L. Rodrigo-Angulo & M. Garzón. 2001. Brain structures and mechanisms involved in the generation of REM sleep. *Sleep Med. Rev.* **5:** 63–77.
5. McGinty, D. & R. Szymusiak. 2001. Brain structures and mechanisms involved in the generation of NREM sleep: focus on the preoptic hypothalamus. *Sleep Med. Rev.* **5:** 323–342.
6. Steiger, A. 2007. Hormonal control of sleep. *Int. J. Sleep Wakefulness* **1:** 9–19.
7. Imeri, L. & M.R. Opp. 2009. How (and why) the immune system makes us sleep. *Nature Rev. Neurosci.* **10:** 199–210.
8. Sassin, J.F., D.C. Parker, J.W. Mace, *et al.* 1969. Human growth hormone release: relation to slow-wave sleep and sleep-walking cycles. *Science* **165:** 513–515.
9. Steiger, A. 2002. Sleep and the hypothalamo-pituitary-adrenocortical system. *Sleep Med. Rev.* **6:** 125–138.
10. Luboshitzky, R., Z. Shen-Orr, A. Ishai & P. Lavie. 2000. Melatonin hypersecretion in male patients with adult-onset idiopathic hypogonadotropic hypogonadism. *Exp. Clin. Endocrinol. Diabetes* **108:** 142–145.
11. Marshall, J.C., A.C. Dalkin, D.J. Haisenleder, *et al.* 1993. GnRH pulses—the regulators of human reproduction. *Trans. Am. Clin. Climatol. Assoc.* **104:** 31–46.
12. Seki, K., T. Uesato, K. Kato & K. Shima. 1985. Twenty-four hour secretory pattern of thyroid-stimulating hormone in hyperprolactinemic women with pituitary microadenoma. *Endocrinol. Jpn.* **32:** 369–373.
13. Schmid, D.A., A. Wichniak, M. Uhr, *et al.* 2006. Changes of sleep architecture, spectral composition of sleep EEG, the nocturnal secretion of cortisol, ACTH, GH, prolactin, melatonin, ghrelin, and leptin, and the DEX-CRH test in depressed patients during treatment with mirtazapine. *Neuropsychopharmacology* **31:** 832–844.
14. Kalleinen, N., P. Polo-Kantola, K. Irjala, *et al.* 2008. 24-hour serum levels of growth hormone, prolactin, and cortisol in pre- and postmenopausal women: the effect of combined estrogen and progestin treatment. *J. Clin. Endocrinol. Metab.* **93:** 1655–1661.
15. Latta, F., R. Leproult, E. Tasali, *et al.* 2005. Sex differences in nocturnal growth hormone and prolactin secretion in healthy older adults: relationships with sleep EEG variables. *Sleep* **28:** 1519–1524.
16. Sakakibara, R., T. Uchiyama, Z. Liu, *et al.* 2005. Nocturnal polyuria with abnormal circadian rhythm of plasma arginine vasopressin in post-stroke patients. *Intern. Med.* **44:** 281–284.
17. Lindow, S.W., A. Newham, M.S. Hendricks, *et al.* 1996. The 24-hour rhythm of oxytocin and beta-endorphin secretion in human pregnancy. *Clin. Endocrinol.* **45:** 443–446.
18. ten Doesschate, T., L.J. Reichert & J.A. Claassen. 2010. Desmopressin for nocturia in the old: an inappropriate treatment due to the high risk of side-effects? *Tijdschr. Gerontol. Geriatr.* **41:** 256–261.
19. Perras, B., U. Wagner, J. Born & H.L. Fehm. 2003. Improvement of sleep and pituitary-adrenal inhibition after subchronic intranasal vasopressin treatment in elderly humans. *J. Clin. Psychopharmacol.* **23:** 35–44.
20. Born, J., C. Kellner, D. Uthgenannt, *et al.* 1992. Vasopressin regulates human sleep by reducing rapid-eye-movement sleep. *Am. J. Physiol.* **262:** E295–E300.
21. Rahm, C., S. Schulz-Juergensen & P. Eggert. 2010. Effects of desmopressin on the sleep of children suffering from enuresis. *Acta Paediatr.* **99:** 1037–1041.
22. Badiu, C., M. Coculescu & M. Moller. 1999. Arginine vasotocin mRNA revealed by in situ hybridization in bovine pineal gland cells. *Cell Tissue Res.* **295:** 225–229.
23. Pavel, S., R. Goldstein, M. Petrescu & M. Popa. 1981. REM sleep induction in prepubertal boys by vasotocin: evidence for the involvement of serotonin containing neurons. *Peptides* **2:** 245–250.

24. Lancel, M., S. Krömer & I.D. Neumann. 2003. Intracerebral oxytocin modulates sleep-wake behaviour in male rats. *Regul. Pept.* **114:** 145–152.

25. Drucker-Colín, R.R., C.W. Spanis, J. Hunyadi, *et al.* 1975. Growth hormone effects on sleep and wakefulness in the rat. *Neuroendocrinology* **18:** 1–8.

26. Wang Z. 1999. Growth hormone deficiency in adults and clinical use of recombinant human growth hormone. *Chin. Med. J.* **112:** 195–201.

27. Kato, Y., H.Y. Hu & M. Sohmiya. 1996. Short-term treatment with different doses of human growth hormone in adult patients with growth hormone deficiency. *Endocr. J.* **43:** 177–183.

28. Roky, R., J.L. Valatx & M. Jouvet. 1993. Effect of prolactin on sleep-wake cycle in the rat. *Neurosci. Lett.* **156:** 117–120.

29. Oba'l, F. Jr., L. Payne, B. Kacsoh, *et al.* 1994. Involvement of prolactin in the REM sleep-promoting activity of systemic vasoactive intestinal peptide (VIP). *Brain Res.* **645:** 143–149.

30. Oba'l, F. Jr., B. Kacsoh, S. Bredow, *et al.* 1997. Sleep in rats rendered chronically hyperprolactinemic with anterior pituitary grafts. *Brain Res.* **755:** 130–136.

31. Bodosi, B., F. Obál, Jr., *et al.* 2000. An ether stressor increases REM sleep in rats: possible role of prolactin. *Amer. J. Physiol. Reg. Integ. Comp. Physiol.* **279:** R1590–R1598.

32. Kok, S.W., F. Roelfsema, S. Overeem, *et al.* 2004. Pulsatile LH release is diminished, whereas FSH secretion is normal, in hypocretin-deficient narcoleptic men. *Am. J. Physiol. Endocrinol. Metab.* **287:** E630–E636.

33. Ismailogullari, S., C. Korkmaz, Y. Peker, *et al.* 2011. Impact of long-term gonadotropin replacement treatment on sleep in men with idiopathic hypogonadotropic hypogonadism. *J. Sex Med.* **8:** 2090–2097.

34. Teran-Pérez, G., Y. Arana-Lechuga, R.O. González-Robles, *et al.* 2010. Polysomnographic features in infants with early diagnosis of congenital hypothyroidism. *Brain Dev.* **32:** 332–337.

35. Steiger, A. 1999. Thyroid gland and sleep. *Acta Med. Austriaca* **26:** 132–133.

36. Feng, P. 2006. The developmental regulation of wake/sleep system. In *Neuroendocrine Correlates of Sleep/Wakefulness.* D.P. Cardinalli & S.R. Pandi-Perumal, Eds.: 1–10. Springer. New York.

37. Vázquez-Palacios, G., S. Retana-Márquez, H. Bonilla-Jaime & J. Velázquez-Moctezuma. 2001. Further definition of the effect of corticosterone on the sleep-wake pattern in the male rat. *Pharmacol. Biochem. Behav.* **70:** 305–310.

38. Vazquez-Palacios, G. & J. Velazquez-Moctezuma. 2000. Effect of electric foot shocks, immobilization, and corticosterone administration on the sleep-wake pattern in the rat. *Physiol. Behav.* **71:** 23–28.

39. Fisher, J.L., J.A. Schwatrzbaum, M. Wrensch & J.L. Wiemels. 2007. Epidemiology of brain tumors. *Neurol. Clin.* **25:** 867–890.

40. Colao, A., L.F. Grasso, R. Pivonello & G. Lombardi. 2011. Therapy of aggressive pituitary tumors. *Expert. Opin. Pharmacother.* **12:** 1561–1570.

41. Sallanon, M., C. Buda, M. Puymartin, *et al.* 1988. Hypophysectomy does not disturb the sleep-waking cycle in the cat. *Neurosci. Lett.* **88:** 173–178.

42. Zhang, J.X., J.-L. Valatx & M. Jouvet. 1988. Hypophysectomy in monosodium glutamate-pretreated rats suppresses paradoxical sleep rebound. *Neurosci. Lett.* **86:** 94–98.

43. Biermasz, N.R., S.D. Joustra, E. Donga, *et al.* 2011. Patients previously treated for nonfunctioning pituitary macroadenomas have disturbed sleep characteristics, circadian movement rhythm, and subjective sleep quality. *J. Clin. Endocrinol. Metab.* **95:** 1524–1532.

44. Dantzer, R., J.C. O'Connor, G.G. Freund, *et al.* 2008. From inflammation to sickness and depression: when the immune system subjugates the brain. *Nature Rev. Neurosci.* **9:** 46–56.

45. Yehuda, S., B. Sredni, R.L. Carasso & D. Kenigsbuch-Sredni. 2009. REM sleep deprivation in rats results in inflammation and interleukin-17 elevation. *J. Interf. Cytok. Res.* **29:** 393–398.

46. Banks, W.A. & M.A. Erickson. 2010. The blood-brain barrier and immune function ans dysfunction. *Neurobiol. Dis.* **37:** 26–32.

47. Marz, P., J.G. Cheng, R.A. Gadient, *et al.* 1998. Sympathetic neurons can produce and respond to interleukin 6. *Proc. Natl. Acad. Sci. USA* **95:** 3251–3256.

48. Breder, C.D., C.A. Dinarello & C.B. Saper. 1988. Interleukin-1 immunoreactive innervation of the human hypothalamus. *Science* **240:** 321–324.

49. Ignatowski, T.A., B.K. Noble, J.R. Wright, *et al.* 1997. Neuronal-associated tumor necrosis factor (TNFα): its role in noradrenergic functioning and modification of its expression following antidepressant drug administration. *J. Neuroimmunol.* **79:** 84–90.

50. Allan, S.M. & N.J. Rothwell. 2001. Cytokines and acute neurodegeneration. *Nature Rev. Neurosci.* **2:** 734–744.

51. Kapsimalis, F., G. Richardson, M.R. Opp & M. Kryger. 2005. Cytokines and normal sleep. *Curr. Opin. Pulm. Med.* **11:** 481–484.

52. Opp, M.R. 2005. Cytokines and sleep. *Sleep Med. Rev.* **9:** 355–364.

53. Capuron, L. & Miller, A.H. 2011. Immune system to brain signaling: neuropsychopharmacological implications. *Pharmacol. Ther.* **130:** 226–238.

54. Petrovsky, N., P. McNair & L.C. Harrison. 1998. Diurnal rhythms of proinflammatory cytokines: regulation by plasma cortisol and therapeutic implications. *Cytokine* **10:** 307–312.

55. Taylor-Gjevre, R., J.A. Gjevre, R. Skomro & B.V. Nair. 2011. Assessment of sleep health in patients with rheumatic disease. *Int. J. Clin. Rheumatol.* **6:** 207–218.

56. Yue, H.J., P.J. Mills, S. Ancoli-Israel, *et al.* 2009. The roles of TNF-α and the soluble TNF receptor I on sleep architecture in OSA. *Sleep Breath* **13:** 263–269.

57. Haack, M., E. Sanchez & J.M. Mullingon. 2007. Elevated inflammatory markers in response to prolonged sleep restriction are associated with increased pain experience in healthy volunteers. *Sleep* **30:** 1145–1152.

58. Vgontzas, A.N., M. Zoumakis, D.A. Papanicolau, *et al.* 2002. Chronic insomnia is associated with a shift of interleukin-6 and tumor necrosis factor secretion from nighttime to daytime. *Metabolism* **51:** 887–892.

59. Irwin, M., J. McClintick, C. Costlow, *et al.* 1996. Partial night sleep deprivation reduces natural killer and cellular

immune responses in humans. *Fed. Am. Soc. Exp. Biol. J.* **10**: 643–653.

60. Heiser, P., B. Dickhaus, W. Schreiber, *et al.* 2000. White blood cells and cortisol after sleep deprivation and recovery sleep in humans. *Eur. Arch. Psychiat. Clin. Neurosci.* **250**: 16–23.

61. Born, J., T. Lange, K. Hansen, *et al.* 1997. Effects of sleep and circadian rhythm on human circulating immune cells. *J. Immunol.* **158**: 4454–4464.

62. Ruiz, F.S., M.L. Andersen, R.C. Martins, *et al.* 2012. Immune alterations after selective rapid eye movement or total sleep deprivation in healthy male volunteers. *Innate Immun* **18**: 44–54.

63. Velazquez-Moctezuma, J., E. Dominguez-Salazar, E. Cortes-Barberena, *et al.* 2004. Differential effects of rapid eye movement sleep deprivation and immobilization stress on blood lymphocyte subsets in rats. *Neuroimmunomodulation* **11**: 261–267.

64. Pandey, A.K. & S.K. Kar. 2011. REM sleep deprivation of rats induces acute phase response in liver. *Biochem. Biophys. Res. Commun.* **410**: 242–246.

65. Brown, R., G. Pang, A.J. Husband & M.G. King. 1989. Suppression of immunity to influenza virus infection in the respiratory tract following sleep disturbance. *Reg. Immunol.* **2**: 321–325.

66. Lange, T., B. Perras, H.L. Fehm & J. Born. 2003. Sleep enhances the human antibody response to hepatitis A vaccination. *Psychosom. Med.* **65**: 831–835.

67. Lange, T., S. Dimitrov, T. Bollinger, *et al.* 2011. Sleep after vaccination boost immunological memory. *J. Immunol.* **187**: 283–290.

68. Everson, C.A. 1995. Functional consequences of sustained sleep deprivation in the rat. *Behav. Brain Res.* **69**: 43–54.

69. Koban, M. & K.L. Swinson. 2005. Chronic REM-sleep deprivation of rats elevates metabolic rate and increases UCP1 gene expression in brown adipose tissue. *Am. J. Physiol. Endocrinol. Metab.* **289**: E68–E74.

70. Jaiswal, M.K. & B.N. Mallick. 2009. Prazosin modulates rapid eye movement sleep deprivation-induced changes in body temperature in rats. *J. Sleep Res.* **18**: 349–356.

71. Everson, C.A. & L.A. Toth. 2000. Systemic bacterial invasion induced by sleep deprivation. *Am. J. Physiol. Regul. Integr. Comp. Physiol.* **278**: R905–R916.

72. Streit, W.J., S.A. Walter & N.A. Pennel. 1999. Reactive microgliosis. *Prog. Neurobiol.* **57**: 563–581.

73. Wisor, J.P., M.A. Schmidt, C. William & B.S. Clegern. 2011. Evidence for neuroinflammatory and microglial changes in the cerebral response to sleep loss. *Sleep* **34**: 261–272.

74. Hsu, J.C., Y.S. Lee, C.N. Chang, *et al.* 2003. Sleep deprivation inhibits expression of NADPH-d and NOS while activating microglia and astroglia in the rat hippocampus. *Cells Tissues Organs* **173**: 242–254.

75. Hsu, J.C., Y.S. Lee, C.N. Chang, *et al.* 2003. Sleep deprivation prior to transient global cerebral ischemia attenuates glial reaction in the rat hippocampal formation. *Brain Res.* **984**: 170–181.

76. Weil, Z.M., G.J. Norman, K. Karelina, *et al.* 2009. Sleep deprivation attenuates inflammatory responses and ischemic cell death. *Exper. Neurol.* **218**: 129–136.

77. Prudlo, J., J. Koenig, S. Ermert & J. Juhász. 2007. Sleep disordered breathing in medically stable patients with myasthenia gravis. *Eur. J. Neurol.* **14**: 321–326.

78. Nicolle, M.W., S. Rask, W.J. Koopman, *et al.* 2006. Sleep apnea in patients with myasthenia gravis. *Neurology* **67**: 140–142.

79. Zamir, G., J. Press, A. Tal & A. Tarasiuk. 1998. Sleep fragmentation in children with juvenile rheumatoid arthritis. *J. Rheumatol.* **25**: 1191–1197.

80. Lopes, M.C., C. Guilleminault, A. Rosa, *et al.* 2008. Delta sleep instability in children with chronic arthritis. *Braz. J. Med. Biol. Res.* **41**: 938–943.

81. Bloom, B.J., J.A. Owens, M. McGuinn, *et al.* 2002. Sleep and its relationship to pain, dysfunction, and disease activity in juvenile rheumatoid arthritis. *J. Rheumatol.* **29**: 169–73.

82. Mahowald, M.W., M.L. Mahowald, S.R. Bundlie & S.R. Ytterberg. 1989. Sleep fragmentation in rheumatoid arthritis. *Arthritis Rheum.* **32**: 974–983.

83. Valencia-Flores, M., M. Resendiz, V.A. Castaño, *et al.* 1999. Objective and subjective sleep disturbances in patients with systemic lupus erythematosus. *Arthritis Rheum.* **42**: 2189–2193.

84. Iaboni, A., D.D. Gladman, M.B. Urowitz & H. Moldofsky. 2004. Disordered sleep, sleepiness, and depression in chronically tired patients with systemic lupus erythematosis. *Sleep* **27**: A327–A328.

85. Jamieson, A.H., C.A. Alford, H.A. Bird, *et al.* 1995. The effect of sleep and nocturnal movement on stiffness, pain, and psychomotor performance in ankylosing spondylitis. *Clin. Exp. Rheumatol.* **13**: 73–78.

86. Gudbjornsson, B., J.E. Broman, J. Hetta & R. Hallgren. 1993. Sleep disturbances in patients with primary Sjogren's syndrome. *Br. J. Rheumatol.* **32**: 1072–1076.

87. Prado, G.F., R.P. Allen, *et al.* 2002. Sleep disruption in systemic sclerosis (scleroderma) patients: clinical and polysomnographic findings. *Sleep Med.* **3**: 341–345.

88. Leigh, T.J., I. Hindmarch, H.A. Bird & V. Wright. 1988. Comparison of sleep in osteoarthritic patients and age and sex matched healthy controls. *Ann. Rheum. Dis.* **47**: 40–42.

89. Jauch-Chara, K., S.M. Schmid, *et al.* 2008. Altered neuroendocrine sleep architecture in patients with type 1 diabetes. *Diabetes Care* **31**: 1183–1188.

90. Ferini-Strambi, L., M. Filippi, V. Martinelli, *et al.* 1994. Nocturnal sleep study in multiple sclerosis: correlations with clinical and brain magnetic resonance imaging findings. *J. Neurol. Sci.* **125**: 194–197.

91. Twilt, M., A.J. Schulten, P. Nicolaas, *et al.* 2006. Facioskeletal changes in children with juvenile idiopathic arthritis. *Ann. Rheum. Dis.* **65**: 823–825.

92. Shoda, N., A. Seichi, K. Takeshita, *et al.* 2009. Sleep apnea in rheumatoid arthritis patients with occipitocervical lesions: the prevalence and associated radiographic features. *Eur. Spine J.* **18**: 905–910.

93. Redlund-Johnell, I. 1988. Upper airway obstruction in patients with rheumatoid arthritis and temporomandibular joint destruction. *Scan. J. Rheumatol.* **17**: 273–279.

94. Yamamoto, J., Y. Okamoto, E. Shibuya, *et al.* 2000. [Obstructive sleep apnea syndrome induced by ossification

of the anterior longitudinal ligament with ankylosing spondylitis]. *Nihon Kokyuki Gakkai Zasshi* **38:** 413–416.

95. Sakurai, N., Y. Koike, Y. Kaneoke, *et al.* 1993. Sleep apnea and palatal myoclonus in a patient with neuro-Behçet syndrome. *Intern. Med.* **32:** 336–339.

96. Turner, G.A., E.E. Lower, B.C. Corser, *et al.* 1997. Sleep apnea in sarcoidosis. *Sarcoidosis Vasc. Diffuse Lung Dis.* **14:** 61–64.

97. Verbraecken, J., E. Hoitsma, C.P. van der Grinten, *et al.* 2004. Sleep disturbances associated with periodic leg movements

in chronic sarcoidosis. *Sarcoidosis Vasc. Diffuse Lung Dis.* **21:** 137–146.

98. Servan, J., F. Marchand, L. Garma, *et al.* 1995. Narcolepsy disclosing neurosarcoidosis. *Rev. Neurol.* **151:** 281–283.

99. Manconi, M., L. Ferini-Strambi, M. Filippi, *et al.* 2008. Multicenter case-control study on restless legs syndrome in multiple sclerosis: the REMS study. *Sleep* **31:** 944–952.

100. Stickgold, R. & M. Walker. 2009. *The Neuroscience of Sleep.* Academic Press. San Diego.

Ann. N.Y. Acad. Sci. ISSN 0077-8923

ANNALS OF THE NEW YORK ACADEMY OF SCIENCES

Issue: *Neuroimmunomodulation in Health and Disease*

Erratum for Ann. N. Y. Acad. Sci. 1257: 125–132

Citi, S., P. Pulimeno & S. Paschoud. 2012. Cingulin, paracingulin, and PLEKHA7: signaling and cytoskeletal adaptors at the apical junctional complex. *Ann. N.Y. Acad. Sci.* **1257:** 125–132.

The incorrect Figure 1 was inadvertently included in the initial online publication of the above-cited article. The figure has been changed to show the correct dotted lines above the scheme of CGN. The correct figure is shown below.

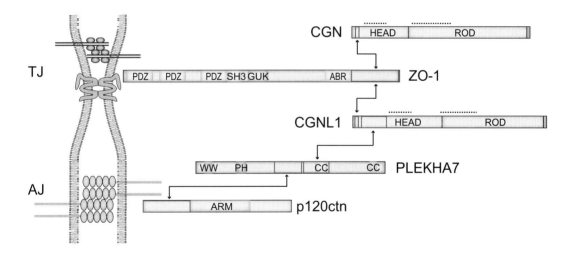

doi: 10.1111/j.1749-6632.2012.06720.x